Development

The bodies of multicellular organisms consist of cells, tissues and organs arranged inside the body in a way characteristic of that species. However, individuals start life either as a single cell (a fertilized egg or spore) or as a small group of cells (a bud) and, through the process of development, the simple embryonic organism is converted into the complex adult. The development of both plants and animals involves increase in the number of cells and changes in the state of cells (i.e. differentiation). In animals, many of the cells of embryos move and change their relative positions during growth but this does not happen in plants because their cell walls are anchored to neighbours. Because of this important difference, we consider plants and animals separately. Many physiological responses of plants are through the growth of cells or organs, so plant development is studied as part of the block on plant physiology (Units 26–31). Units 11–15 deal almost exclusively with the development of animals and lead on to the block on animal physiology (Units 16–25).

Units 11–15 are of very different lengths. Unit 11, which is an introductory survey of the processes of development, is short. Units 12 and 13, combined into a single text, deal with the subcellular changes associated with development and make up two weeks' average study time. Unit 14 is fairly long, dealing with processes of pattern formation, and Unit 15 is short, discussing the origin of patterns of development in the life of individuals. The basic theme of all these Units is regulation and control but sources of energy and structure–function relationships are, of course, important to developing organisms.

The TV programmes contribute in an important way to the teaching of developmental biology because they illustrate the dramatic changes in shape that occur in embryos. They also show some techniques used to investigate the basic mechanisms and include animations explaining basic concepts. We advise you to try to watch them in colour.

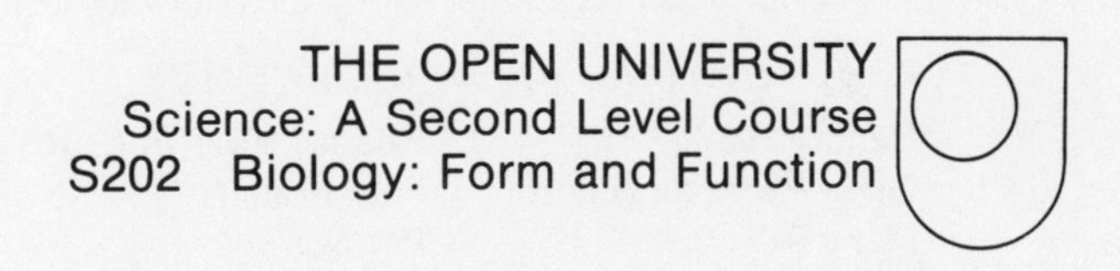

Development I

unit 11

Development: The Component Processes

units 12 and 13

Cellular Differentiation

Prepared by the S202 Course Team

The S202 Course Team

This Course has been prepared by the following team:

Peggy Varley (*Chairman and General Editor*)

Hendrik Ball (*BBC*)
Gerry Bearman (*Editor*)
Mary Bell
Eric Bowers
Bob Burgoyne
Ian Calvert
Norman Cohen
Peter Cole (*BBC*)
Bob Cordell
Baz East (*Illustrator*)
Vic Finlayson
Anna Furth
Denis Gartside (*BBC*)
Lindsay Haddon
Robin Harding
Stephen Hurry
David Kerrison (*BBC*)
Aileen Llewellyn (*BBC*)
Pam Mullins
Pat Murphy
Seán Murphy
Pam Owen (*Illustrator*)
Phil Parker (*Course Coordinator*)
Ros Porter (*Designer*)
Rob Ransom
Irene Ridge
Steven Rose
Ian Rosenbloom (*BBC*)
Jacqueline Stewart (*Course Editor*)
Mike Stewart
Jeff Thomas
David Tillotson (*Editor*)
Charles Turner
Sue Turner

The Open University Press,
Walton Hall, Milton Keynes.

First published 1981.

Designed by the Graphic Design Group of the Open University.

Typeset by Santype International Limited, Salisbury, Wilts, and printed by Linneys.

ISBN 0 355 16034 4

This text forms part of an Open University course. The complete list of Units in the Course is printed at the end of this text.

For general availability of supporting material referred to in this text please write to: Open University Educational Enterprises Limited, 12 Cofferidge Close, Stony Stratford, Milton Keynes, MK11 1BY, Great Britain.

Further information on Open University courses may be obtained from the Admissions Office, The Open University, P.O. Box 48, Walton Hall, Milton Keynes, MK7 6AB.

1.1

unit 11

Development: The Component Processes

Contents

TABLE A Scientific terms and principles introduced in Unit 11

Assumed knowledge†	Introduced in an earlier Unit	Unit	Introduced or developed in this Unit	Page
cell differentiation	antibody	5	allele	5
chromosome[3, 8]	antigen	5	animal pole*	23
DNA	blastocoel	1	animal–vegetal axis*	23
dominant[7]	blastula	1	autosome	5
gamete	ectoderm	1	blastocoel*	11
gene	endoderm	1	blastomere*	11
genotype	gastrula	1	blastula*	11
haemoglobin	mesoderm	1	cell differentiation*	12
heterozygous	placental mammals	3	cell movement*	12
homeostasis[1]			cell/tissue grafting*	15
homologous pairs of chromosomes			cleavage*	10
homozygous			clone*	6
in vitro			commitment*	18
in vivo			competence*	20
macromolecule			cytogenetics	6
meiosis[4]			determination*	19
Mendel's rules[6]			developmental biology*	3
metabolism			diploid*	4
mitosis			ectoderm*	11
morphogenesis			embryonic induction*	19
mutation			embryonic inductor*	19
nucleotide bases[5]			endoderm*	11
ovum			gametogenesis*	9
phenotype			gastrula*	11
protein			gastrulation*	11
recessive[2]			growth*	12
RNA			haploid*	4
somatic cell			heterogametic	5
spermatozoa (sperm)			homogametic	5
TCA cycle			mesoderm*	11
zygote			metaplasia	22
			model system*	13
			morphogenesis*	12
			nuclear transfer*	15
			oogenesis*	10
			parthenogenesis*	10
			pattern formation	12
			polarity (egg/embryo)*	23
			reciprocal hybrids*	21
			regulation*	20
			sex chromosomes	5
			somatic cell*	6
			somatic mutation*	6
			species hybrids*	21
			spermatogenesis*	9
			temporal aspects of development*	18
			teratogen	8
			tissue culture*	14
			vegetal pole*	23

* These terms must be thoroughly understood—see Objective 1.

† Most of these terms are explained in the Science Foundation Course. Those that are of particular importance to the understanding of this text appear with a superscript number and have a full reference at the end of the Unit.

Study guide for Unit 11

This Unit serves as an introduction to the Development Block. As such, it attempts to mention briefly some of the key processes, concepts and experimental techniques dealt with in more detail in following Units (12–15). This Unit is short and should not take long to study. Those of you who have studied S101 (Units 19 and 25) should find Section 2 of this Unit (which deals with basic genetics) familiar ground and thus easily and rapidly studied. If you have not studied S101 or genetics elsewhere, Section 2 will give you an adequate background to fulfil the relevant Objectives of this Unit and cope with parts of this Block where a knowledge of genetics is useful.

A word about television. Development is often a highly visual and dynamic process dealing as it does with change. It is therefore particularly important that you watch the TV programmes associated with Units 11–15 as it is only through them that we can hope to convey the dynamic, and often strikingly beautiful, aspects of development.

1 Introduction

Up till now a central theme of this Course has been the relationship between structure and function: how the structure of the cell and the parts within it are fitted for their functions and how function depends on structure. We have also been at pains to emphasize the dynamic aspects of cell biology. However, the majority of those dynamic events that occur in cells, at the metabolic or organelle level, are organized to maintain the *status quo*, keep the cell in dynamic equilibrium, maintain what is termed homeostasis[1]. Later in this Course you will see that at the physiological level, systems or organs interact to maintain homeostasis; structures once again appear able to adapt appropriately to conditions, thus preventing overall change. There are, however, two major areas of biology that are primarily concerned with overall change.

One of these is *evolutionary biology*, where the very name means change. Here the changes are very slow, organisms often evolving over millions of years. Again the driving force is a need to fit structure to function, and natural selection operates to favour those organisms best fitted to their surroundings at any particular time. At the level of the individual organism usually no change is obvious within the lifespan of an observer. This is because the changes are extremely slow and subtle and, more importantly, the changes are relative and whole populations of organisms must be compared. The other area of biology where change is paramount is *developmental biology*. Here the changes are much more rapid and obvious: the transformation from egg to adult is indeed dramatic. It is to developmental biology that we now transfer.

developmental biology*

All adult multicellular organisms, whether they spring from egg, bud or seed, are the products of a developmental history during which the shape, size, anatomy, physiology and chemistry of the individual undergo profound changes. The nature of these changes and the ways in which they are controlled are the primary concern of developmental biologists.

Clearly there must be as many kinds of developmental history as there are kinds of organism. Fortunately these developmental histories have much in common, and it is to the fundamental common factors in development, to the problems they pose and to the way these problems are tackled that this Unit will introduce you. Many of these features are elaborated in Units 12–15.

However, the scientific study of development relies upon the fact that any two animals or plants of the same species will develop by the same processes and obey the same laws. Clearly, the features that members of the same species have in common, and those in which they may differ, are crucial to experimental work. We therefore begin by outlining some of the genetic background to development (Section 2). This is followed by a consideration of the scientific and social relevance of studies on developmental biology (Section 3). The rest of the Unit is devoted to a brief consideration of the component processes of development and some guidelines as to how they are studied.

2 The genetic background

Much of this Section goes over basic genetics. We hope that you will appreciate how a knowledge of genetics is valuable in understanding development.

It is a matter of common experience that 'like begets like'. This is generally true within a species, so that, with rare exceptions, parents and offspring belong to the same species. It is also true that offspring resemble their parents more closely than they do more distantly related members of their own species. However, in populations in which the production of offspring depends on the fusion of male and female gametes, as the parents must differ from each other, if only in sex, the resemblance cannot approach identity to both parents. In practice, we find that the phenotype of an organism, if compared with the phenotype of its parents, shows the following (as exemplified in Table 1):

1 Some features in which it is indistinguishable from either parent.

2 Some features in which it is indistinguishable from one, but differs from the other.

3 Some features in which it differs from both.

TABLE 1 Resemblance between parents and offspring

Individuals	Sex*	Height† in m at age 25	Blood group (ABO)	Haemoglobin	Blood pressure‡ at age 40 (kPa)	PTC§ threshold ($mg\,l^{-1}$)	Blood group‖	Blood-clotting¶ time (mins)
Parents								
Mr. LKJ	Male (♂)	1.90	B	HbA	22/13	0.01	Rh^-	7.0
Mrs. LKJ	Female (♀)	1.82	O	HbA	19/12	140	Rh^-	8.2
Offspring								
Mrs. MO (nee J)	♀	1.76	B	HbA	19/11	0.02	Rh^-	7.3
Mr. DJ	♂	1.92	B	HbA	19/12	280	Rh^-	6.9
Mr. KJ	♂	1.89	O	HbA	20/13	0.02	Rh^-	810

You should note that:

* Less than 1 per cent of adult human subjects are biologically ambiguous as to sex.

† Height is sensitive to the experience of the individual in nutritional and other respects, but with 'optimum' nutrition offspring tend to lie between the mean parental height and the population mean, but do not always do so.

‡ Blood pressure varies in any one individual from time to time. The figures given in Table 1 should therefore be treated with caution. Mr. LKJ's is a little on the high side for a male of his age.

§ The minimum concentration of PTC (phenylthiocarbamide) that can be tasted varies widely in human populations. 'Non-tasters', that is those who cannot detect low concentrations, are homozygous for a recessive[2] gene.

‖ Both parents are homozygous; the children follow suit.

¶ Blood-clotting times can be measured in different ways and the time depends upon both the method and the care with which it is used. However, Mr. KJ is wildly beyond the normal range and is, in fact, a haemophiliac. (Haemophilia is a disease in which the blood fails to clot.)

In principle, it is possible that both inheritance (the resemblance) and variation (the differences) could be due to similarities and differences in the life experience of each individual. The environment could determine what sort of organism the individual becomes. This view in its extreme form is quite untenable, though its rejection does not mean that our phenotypes are independent of the environment. As you will see, some properties of organisms are almost wholly insensitive to environmental influence while others are more or less sensitive.

As an example consider human monozygotic twins which have as much in common as it is easily possible to have. A pair of monozygotic twins are derived from a single fertilized egg and have shared the same uterine environment; both will often spend their childhood in the same general physical environment. Yet though their resemblances can be striking and in many chemical features they will be indistinguishable, they do differ visibly, sometimes in quite dramatic ways.

Sexual or biparental inheritance endows the fertilized egg with two broadly equivalent sets of chromosomes, one set from each parent.[3] Chromosomes can often be distinguished from one another by size and shape and it is found that each set can be matched against the other in homologous pairs. The single set is called a *haploid* set, the double set a *diploid* set. During mitotic nuclear division each

haploid* **diploid***

chromosome yields two apparently identical daughter chromosomes, of which each daughter cell receives one. All the cells of an organism thus contain a replica of all the chromosomes contributed by the parents.

The equivalence of the haploid sets contributed by the parents is subject to two kinds of qualification. The first concerns the so-called *sex chromosomes*. It is usual for one sex to be characterized by the possession of two X-chromosomes, the other by one X and one Y. The former, in producing by means of meiosis[4] haploid sets for its gametes must give each set one X and is therefore called *homogametic*; the latter gives half its gametes an X and half a Y and is therefore called *heterogametic*. Those chromosomes which are not sex chromosomes are known as *autosomes*.

sex chromosomes

homogametic

heterogametic

autosome

The second qualification emerges from a detailed study of the actual genetic information in homologous ('sister') chromosomes. This takes the form of serially arranged lengths of DNA, or genes, which lie end to end along the chromosome. We might expect homologous chromosomes to carry homologous genes, and in general they do. However, the actual genetic information carried by the gene may differ between the two homologues because of differences in the sequence of nucleotide bases[5]; we then say that two *alleles* (or alternative forms of the gene) are present in the cell. Where the two genes are identical, or at least indistinguishable, the cell and the whole organism of which it is part is said to be *homozygous* for that gene. Where the two alleles are different we say the organism is *heterozygous* for that gene.

allele

In the production of gametes, nuclear divisions, called meiosis, lead to the formation of haploid nuclei. The haploid sets provided for the gametes are, however, randomly selected from chromosomes of paternal and maternal origin. Furthermore, during meiosis a process of exchange of genetic material between homologous chromosomes may occur, so that any chromosome in the gamete is itself a mixture of maternally and paternally derived pieces. The consequence is that each gamete will have one representative of each gene, but that overall these will be chosen at random from those ultimately derived from the organism's parents*.

The sum of the genes present in an organism will thus consist of many homozygous pairs of alleles and many heterozygous ones. Where a gene is represented by a pair of different alleles, they are often both effective in contributing to the phenotype. An example would be the ABO blood group system. An individual, heterozygous in possessing both A and B alleles, shows both A and B substances on its cells. In other cases one member of a heterozygous pair may fail to make its presence felt. It is then designated as *recessive* to its *dominant* partner. For example, certain human beings lack pigments in the skin, a condition known as *albinism*. Most albinos are the children of non-albino parents.

□ How can you explain this in terms of recessive versus dominant alleles? (Bear in mind Mendel's rules.[6])

■ The father has one 'albino' allele *a* plus one 'non-albino' allele *A*. It follows that half the gametes he produces will carry the albino allele, *a*. The mother is in the same position. Assuming that the non-albino allele is dominant[7] to the albino we would expect the result of a mating between the two parents to be as shown in the margin. So, for the offspring we expect one-quarter should have two 'non-albino' alleles *A*//*A* and be *non-albino in appearance*, one-half should have one albino allele only *A*//*a* (like their parents) and be *non-albino in appearance* and one-quarter should have two albino alleles *a*//*a* and be *albino in appearance*. Hence some (one-quarter) of the offspring could be albino from non-albino parents.

parents $\frac{A}{a} \times \frac{A}{a}$

↓

GENOTYPE OF SPERMS

GENOTYPE OF EGGS	*A*	*a*
A	$\frac{A}{A}$	$\frac{A}{a}$
a	$\frac{a}{A}$	$\frac{a}{a}$

Mendelian inheritance implies very considerable accuracy in the process of replicating genes during cell proliferation. Yet altered alleles do occur 'spontaneously' rather rarely. They are known as gene mutations and can take the form of a change in part of the DNA sequence constituting an allele, that is, a change in the nucleo-

* The haploid number, that is the number of chromosomes per haploid (gamete) cell, for a species is often given the abbreviation *n*. The diploid ($2n$) numbers (number of chromosomes in a somatic cell) for some organisms favoured by geneticists are:

fruit-fly (*Drosophila melanogaster*)	$2n$	8
human (*Homo sapiens*)	$2n$	46
mouse (*Mus musculus*)	$2n$	40
maize (*Zea mays*)	$2n$	20

tide base sequence, by loss or substitution. (Spontaneous mutation rates in humans are estimated, for different genes, to lie between 1 in 10000 and 1 in 100000 per allele per gamete. Mutation rates can be raised by high temperature, some chemical agents and by ionizing radiation.) It is characteristic of mutant alleles that their own subsequent replication is also very accurate. Mutations in cells that are destined to give rise to tissues of the body (so-called *somatic cells*) are known as *somatic mutations*. They will often pass unnoticed. Occasionally the descendants of a somatic cell that has mutated are seen as patches of aberrant skin with different colour or hair form from the rest of the body. Somatic mutation is thought to be responsible for some cases in humans in which the eyes differ in colour or in which one segment of one iris is different in colour from the rest.

somatic cell*

somatic mutation*

ITQ 1 Will the stage of development at which a somatic mutation occurs influence its final effect, as seen in the adult organism?

Answers to ITQs begin on p. 24

Modern genetics has thus provided an understanding of what lies behind both inheritance and variation in animals and plants. DNA contains, in the sequence of the nucleotide bases, information that specifies the amino acid sequence in proteins, and so cells with the same complement of DNA molecules have the same repertoire of possible proteins that they could synthesize. Because the DNA molecule can be copied with great precision to produce two 'daughter' molecules with the same nucleotide base sequence as the 'parent' molecule, if we ignore mutation any *clone* of cells (a 'clone' is a group of cells that are all descended by mitotic cell division from the same ancestor cell) will have the same genetic information and hence the same range of proteins available to all its members. This limits the effect of the environment of the cell to controlling or selecting which part of the genetic information present in it should be used (transcribed and translated) at any one time. In development, however, such selection presents a critical problem.

clone*

Thus, if we consider the incentives for studying development, an important place must be given to the need to understand how genetic information is put to use in different ways in the different cells of the organism.

Objectives and SAQs for Section 2

Now that you have completed this Section, you should be able to:

★ understand relevant terms and principles in Table A.

★ evaluate evidence attributing a role to genetic information (genotype) in development.

To test your understanding of this Section, try the following SAQs.

SAQ 1 (*Objectives 1 and 2*) A human ovum containing one X-chromosome is fertilized by a sperm containing one Y-chromosome. Will the resultant child be male or female?

SAQ 2 (*Objectives 1 and 2*) A red, female flower is fertilized by pollen from a red flower taken from another plant of the same species. Some of the seeds thus produced give rise to plants with white flowers, some to plants with red flowers. Ruling out mutation, how could this arise in a simple way?

3 Incentives for study

One obvious incentive for studying development is to increase our basic knowledge of this important branch of biology. However, in addition there are several possible applications of any knowledge gained.

3.1 Completing the cytogenetic scheme

The 50 years between 1860 and 1910 were a golden age for biology. Old misconceptions disappeared as evolutionary theory was vindicated and as belief in 'spontaneous generation' finally disappeared. Above all, cell theory approached its present position when the fundamental nature of the mitotic and meiotic cycles and of fertilization became clear. When the first links between genetics and chromosome behaviour were established, giving rise to the notion that genes were within chromosomes[8], a coherent biology of the multicellular organism was in the making. But over it all hung an apparent paradox, which is only now yielding to direct investigation.

In its simplest form, *cytogenetics* (the combined study of genetics and cell structure) sees a higher plant or animal as a clone of cells descended from the fertilized egg through mitotic nuclear divisions. Each cell contains the same genetic information as the others and the same as the zygote had at fertilization. Although important exceptions to this general statement are known (Units 12 and 13), what matters is that it is broadly true in many cases. Yet the adult organism consists of cells that are obviously different from each other in size, shape, physiological activity and chemical constitution. The differences are essential to the varied contributions that cells make to the economy of the whole organism.

cytogenetics

Thus the cytogenetic scheme cannot be complete until the gap between the genotype created at fertilization and the phenotype visible in later life has been bridged by an understanding of the development of the organism, a development that involves the creation and maintenance of significant differences between cells of identical genetic origin. This challenge has been recognized for many years and many approaches to it have been suggested. As we shall see (Units 12 and 13), the era of molecular biology has given a new impetus to work in this field, but we still seek general solutions to the problems of development.

3.2 Congenital defect

No one knows with any certainty the prevalence of prenatal death in humans. The existence of any embryo dying before the 'first missed period' will not be suspected and deaths before 4 weeks may often go unnoticed by the mother. The balance of opinion is that more than 30 per cent of all fertilized human eggs fail to come to normal birth. The true figure could be much higher.

It is perhaps not surprising that of those who do come to live birth about 4 per cent suffer from some, often very slight but in some cases severe, congenital defect. (Congenital only means 'present at birth'; it does not imply anything about the origin of the defect.) This may be considered to be the tip of the iceberg of prenatal disease, but in terms of human suffering it is a serious tip.

The congenital defects that we see are of many kinds—failure to achieve normal anatomical structure, chemical deficiency, physiological accident. If we wish to avert them we must clearly know something of their cause. Are they wholly genetic? Do they arise wholly from the conditions of pregnancy? Or are the defects the result of a complex of genetic and environmental influences? There are many clues from which to judge. Let us take some examples.

1 Soon after thalidomide was introduced as a popular tranquillizer it became apparent that many women who had taken it in early pregnancy later gave birth to deformed children. Thalidomide was thus suspected of 'causing' deformity (Figure 1). Before the matter was cleared up an alternative hypothesis was pre-

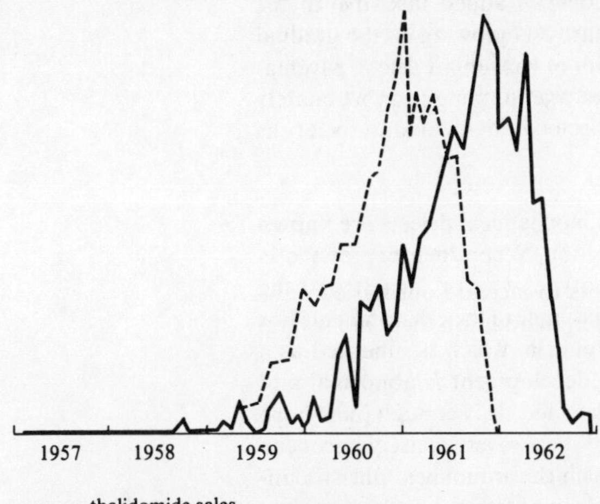

FIGURE 1 The relationship between malformations of the thalidomide type and the sales of thalidomide.

sented. This was that thalidomide, far from causing deformity, rescued from an otherwise certain death embryos that were deformed for other reasons. Two arguments led to the conviction of thalidomide as a *teratogen* (an agent producing deformity in embryos). A careful retrospective study of the available evidence suggested that nearly all mothers who had taken the drug early in pregnancy produced affected children. It seemed impossible that nearly all embryos were doomed to have the kind of deformities associated with thalidomide. Second, direct experiment on the embryos of some other animals showed unequivocally that thalidomide damaged them in the same way. It is therefore reasonable to view thalidomide as a powerful poison for human embryos, and which also damages embryos of many genotypes. Of course, the genotype of the embryo, or of the mother, may affect the severity or the kind of damage to some extent, but in practice thalidomide alone is treated as the culprit.

teratogen

2 A number of other drugs are associated with infant misery of a less severe and less constant kind. The smoking of tobacco or the drinking of alcohol by the pregnant mother increases the risk of a sick infant. The risk, however, does not approach 100 per cent even for heavy smokers or drinkers. It is thus possible that whether or not an infant suffers in these cases is partly a matter of its mother's, or its own, genetically determined susceptibility to the poisons concerned.

3 Some diseases of the new-born are products of an interaction between mother and fetus which has a known genetic basis. For example, as well as the ABO blood groups in humans there are several others including the rhesus (Rh) group. Those individuals with the rhesus substance on their red blood cells are called rhesus positive (Rh^+), those without are rhesus negative (Rh^-). Because the Rh^+ genotype is dominant, an Rh^+ individual can arise from one Rh^- and one Rh^+ parent, or from two Rh^+ parents. If a Rh^+ fetus is in the uterus of a Rh^- mother, complications can arise if fetal cells somehow enter the maternal bloodstream, as the presence of Rh^+ antigens from the fetus in the bloodstream of the mother can cause the mother to produce antibodies against the Rh^+ factor. Such antibodies can cross back through the placenta into the fetus and then attach to and damage its own red blood cells. (The most likely time for the initial passage of fetal cells into the maternal bloodstream is at birth, if placental bleeding occurs. If as a result the mother produces antibodies against the rhesus substance then any Rh^+ fetus of *subsequent* pregnancies might be at risk.) When such a Rh^+ fetus is 'attacked' by antibodies produced in a Rh^- mother, the limitations of the theory of a simple genetic–versus–environmental cause of disease are readily seen. Here, the mother is clearly part of the environment of the fetus, yet the attack will occur only if the genotype of both is appropriate, and indeed is most unlikely to be serious unless the mother has had previous Rh^+ children, as only then will the mother produce much anti-Rh^+ antibody in response to secondary contact with the Rh^+ antigen. In recent years it has proved possible to detect the presence of a Rh^+ fetus while still *in utero* in a Rh^- mother and prevent disastrous outcomes, for example, by changing completely the blood of the unborn infant or by preventing the mother from forming anti-Rh^+ antibody. These innovations also depended on various technical advances coupled with an understanding of the rhesus blood groups and their genetic interrelationships.

The hope of controlling human suffering provides an added incentive in the approach to development and genetics. This incentive is growing as the gradual conquest of infectious diseases leaves the residuum of congenital defect, psychiatric disorder, and the degenerative diseases of old age looming proportionately ever larger in the medical burdens borne by the affected individuals and society as a whole.

4 A number of malformations and a number of biochemical defects are known to be 'inherited', quite strictly, in a Mendelian fashion. When their expression is uniform in all children with the appropriate genotype, we can suggest that the disease is 'caused' genetically. Even so it is wrong to ignore the environment. For example, there is a rare disease called phenylketonuria, which is inherited as a single recessive gene. If left untreated, post-natal development is abnormal and severe mental retardation can result. However, if detected early enough (new-born babies are routinely tested in the United Kingdom) these severe consequences can be avoided by feeding the child on a diet from which the amino acid phenylalanine is severely restricted, as it is abnormal metabolism of this naturally occurring substance that ultimately produces the symptoms.

5 There are two main schools of thought concerning certain mental illnesses. One school attributes cause to the social and familial conditions of the individual, the other to some biochemical lesion in the individual's metabolism. But can one rather than the other be said to be the 'cause', or is there indeed some other possibility? For instance, the social factors might produce the biochemical lesion, or the biochemical lesion may make the individual ill at ease with his surroundings, or the biochemical lesion, if indeed it exists, may be a secondary effect of the illness and of no relevance to the so-called abnormal behaviour.

Examples 1–5 above illustrate the problems associated with attributing 'cause' in biology. It is probably acceptable to attribute the cause of, say, a rapid chemical reaction to one particular enzyme but when dealing with complex systems it is difficult to attribute cause to just one element in that system. Frequently, the 'cause' depends on a complex interaction between genetic and environmental factors.

3.3 Ageing

The processes of senescence are of obvious interest to human societies. We are so familiar with them that we may forget that they demand explanation. It is not self-evident that a clone of cells should become increasingly prone to death with time. The failures of the homeostatic mechanisms of the body which become apparent with time are as much part of its developmental history as was the establishment of these mechanisms early in life.

3.4 Cancer

Cancers consist of cells that break some of the rules that other cells obey: they divide in an uncontrolled fashion; they invade healthy tissues; they may move and colonize parts of the body far from their site of origin. Embryonic cells do such things, but in a controlled way. It is often suggested that our investigations of embryonic control systems will throw light on the behaviour of cancer cells.

4 Key episodes in development

This Section outlines the major events and component processes common to the development of many organisms. As such it is important for the Block as a whole.

Although the cycle of development from zygote to sexually mature adult is often complicated to describe and difficult to analyse, it is helpful to single out some of the key episodes and kinds of process that contribute to it.

4.1 Key episodes in the development of multicellular animals

GAMETOGENESIS (THE FORMATION OF GAMETES)

gametogenesis*

In order to produce a diploid ($2n$) zygote, fertilization must involve two haploid (n) gametes, that is a spermatozoon (a sperm) and an ovum, produced by meiotic division from diploid cells.

1 *Spermatogenesis (the production of spermatozoa)* Spermatozoa are almost always small, haploid, motile cells with very little cytoplasm (Figure 2).

spermatogenesis*

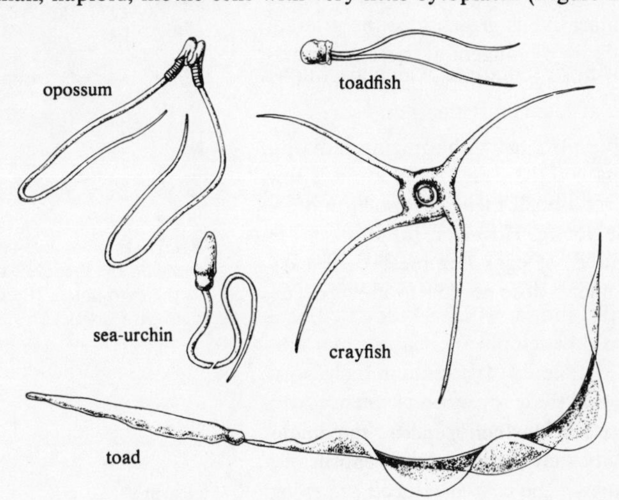

FIGURE 2 Variations in the shape of spermatozoa.

Spermatozoa do, however, contain extranuclear elements which are capable of replication—the centrioles (structures involved in mitosis, etc.) and mitochondria. Normally all the products of male meiosis become sperm, a so-called primary spermatocyte giving rise to four spermatozoa. In mammals and many other animals two of each four spermatozoa contain an X-chromosome, two a Y, and upon the chromosome constitution of the sperm depends the sex of the zygote to which it contributes.

2 *Oogenesis* (*the production of ova*) Ova (eggs) are always large, and in some species vast, cells. The diameters of some of the eggs most studied in developmental work are given below. In these species the egg is approximately spherical:

oogenesis*

man (*Homo sapiens*)	120 μm
mouse (*Mus musculus*)	70 μm
S. African frog (*Xenopus laevis*)	1.2 mm
chicken (*Gallus domesticus*)	30 mm

But within a species, egg size may be variable. Most somatic cells are much smaller, for example, a mammalian liver cell is about 20 μm in diameter.

The size of eggs is partly a reflection of the cytoplasmically stored food resources, or yolk, upon which early embryonic nutrition depends. Meiosis provides one viable egg cell nucleus whereas the other products die in small cells known as polar bodies. Meiosis, in most species, does not proceed to completion until after fertilization. Females are heterogametic in birds, moths and butterflies, and many other animals. In these instances, the sex of the zygote depends upon the constitution of the egg nucleus. Although many eggs are spherical in shape they are polarized in internal structure (e.g. in the distribution of the yolk and the position of the nucleus) and may show bilateral symmetry. We shall see later (Unit 15) that the organization of the egg is of critical importance in early development.

FERTILIZATION

In species with small eggs it is usual for only one spermatozoon to enter the egg (monospermy) (Figure 3); large eggs may be entered by many sperm (polyspermy) of which only one contributes a nucleus (and its chromosomes) to the zygote.

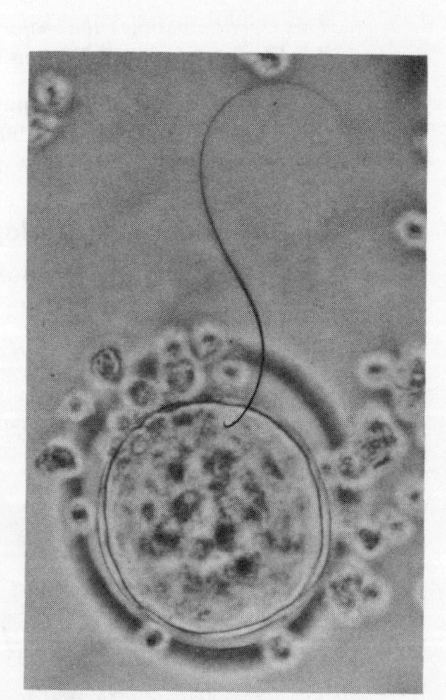

FIGURE 3 A live fertilized rat egg, showing the 'head' of the spermatozoon in the cytoplasm. (By courtesy of Prof. A. Parkes.)

In addition to the bringing together of male and female chromosomes, fertilization activates the egg thus stimulating it to (a) complete meiosis when necessary, (b) inhibit the entry of further spermatozoa in monospermic eggs, (c) undertake some reorganization of cytoplasmic structure (or so it seems in some species, Unit 15), and (d) divide as a prelude to a period of active, mitotic, cell division. Fertilization, however, may not be necessary for the development of the egg. *Parthenogenesis*—development without fertilization—always occurs in species without males (e.g. some lizards, rotifers, and water fleas), sometimes in other species (e.g. wasps, turkeys, aphids) and can be artificially provoked in yet others (e.g. frogs).

ITQ 2 How would you explain the following observations which are known after their discoverer as the Hertwig effect?

Eggs fertilized by spermatozoa that have been treated with ionizing radiation which can damage DNA, (a) develop normally if the dose of radiation is low. (b) As the dose administered to the sperm is raised, a higher proportion of the fertilized eggs develop abnormally and die young. However, (c) above a certain dose this result no longer holds and most eggs complete their early development fairly normally. (d) At a still higher dose no development occurs at all.

CLEAVAGE

Soon after fertilization (or activation in the case of parthenogenesis) the single-celled zygote enters a phase of mitotic activity that leads to the creation of a population of cells. The process is called *cleavage*, the cells produced are called

cleavage*

blastomeres and the stage of development achieved is known as a *blastula* (*pl.* blastulae). The blastula is a hollow ball of cells or a distorted version of this (Figure 4); the cavity is the *blastocoel*, and there may be as few as 32 up to as

blastomere* **blastula***

blastocoel*

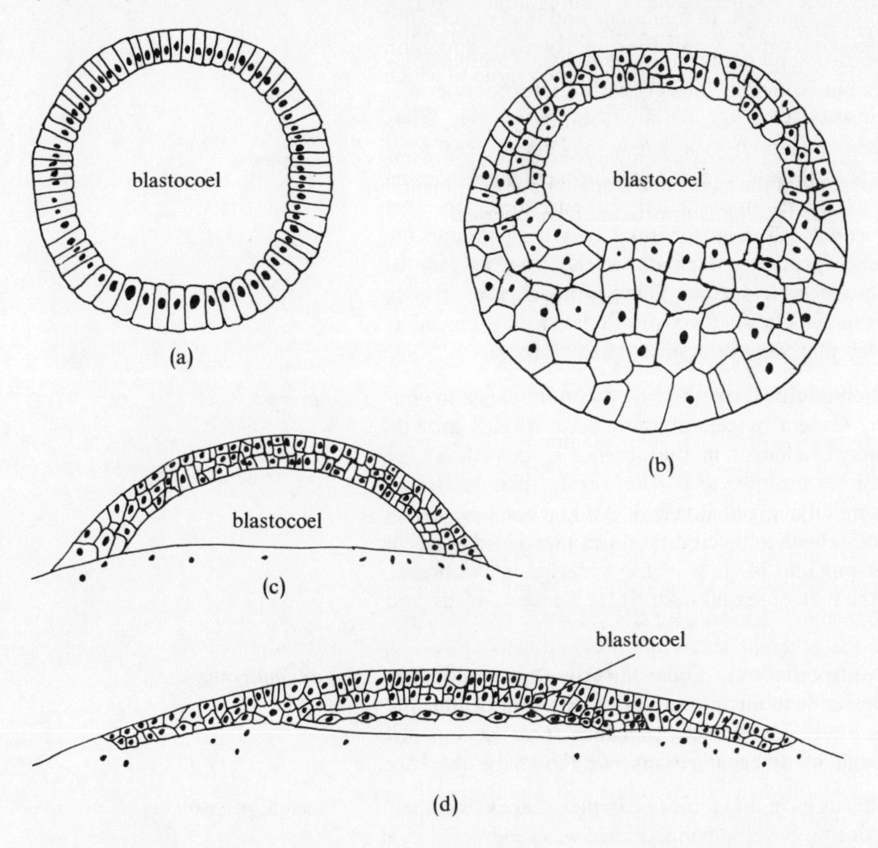

FIGURE 4 Diagrammatic comparison of blastulae of (a) an echinoderm, (b) a frog, (c) a fish, and (d) a bird.

many as several million cells, according to species. Generally, cleavage sees rather small changes in the relative positions of egg substance, at least in insects and amphibians; it serves to partition the substance of the egg into cells.

GASTRULATION

During development cell movement can often be observed. The first major movement to occur is during *gastrulation* which creates an embryo termed a *gastrula*, having an anatomy foreshadowing that of the adult. Layers and groups of cells now come to occupy approximately the positions their descendants will have in the adult organism. The process of modelling the form of the whole, and of its parts or organs, is prolonged. The great movements of gastrulation are followed by less striking minor adjustments. At the end of gastrulation in higher animals it is possible to recognize three major populations of cells, an outer *ectoderm*, an inner *endoderm* and, between them, the *mesoderm* (Figure 5).

gastrulation* **gastrula***

ectoderm*
endoderm* **mesoderm***

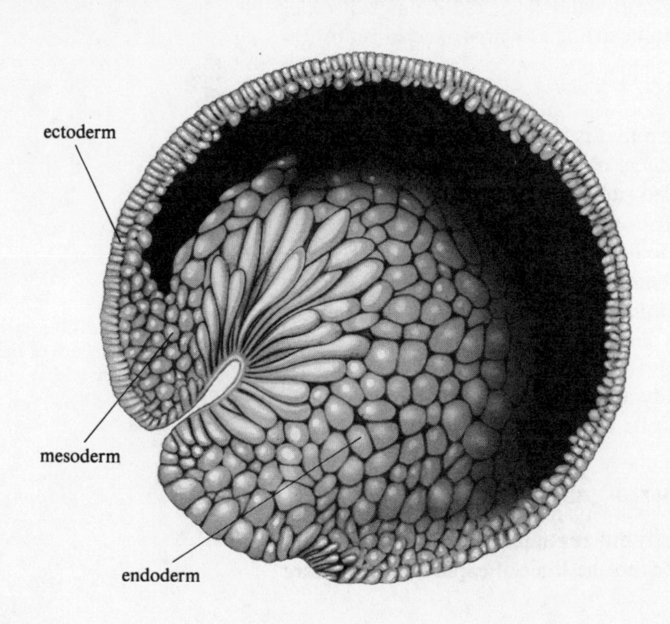

FIGURE 5 Schematized section through an amphibian gastrula.

4.2 The component processes of development of multicellular animals

We have now reviewed some of the key episodes in animal development in the temporal sequence in which they occur: gametogenesis; fertilization; cleavage; gastrulation. There are, of course, later episodes that we have not dealt with. Those we have introduced are all early and common to virtually all multicellular organisms—or at least common in principle though with some considerable modifications (e.g. in certain insects 'cleavage' is modified, Unit 14). What developmental biologists are often concerned with is *how* they occur. To understand *how*, it is convenient to regard these and other developmental episodes as consisting of certain component processes. In particular, there are three processes that taken together make up virtually all developmental episodes—though the manner and extent to which each process contributes varies from episode to episode and to some degree from species to species. These processes are: growth, cell differentiation, and morphogenesis. The last two we will deal with in considerable detail in Units 12–14, but for now we will briefly define all three.

1 *Growth* Eggs are smaller than adults. Growth must therefore occur to convert the former into the latter. Generally, cell division accompanies growth. However, sometimes cell division can occur in the absence of growth, as for example during cleavage. Sometimes the reverse is true; certain plant cells can grow enormously without accompanying cell division. Also, when not totally uniform throughout an organism, growth and cell division can lead to changes in shape. Growth may continue throughout life (e.g. in most marine invertebrates, and fishes) or may cease about the time of sexual maturity (e.g. insects, birds, and mammals).

growth*

2 *Cell differentiation* (*or cytodifferentiation*) Cells of early embryos do not closely resemble, in appearance, in function, or in chemical composition, the differentiated tissue cells of the adult. The process by which these various cell types arise is termed *cell differentiation* or *cytodifferentiation* (Units 12 and 13).

cell differentiation*

3 *Morphogenesis* Adult organisms have much more complex shapes than their eggs. The generation of shape during development, termed *morphogenesis*, is a highly complex and often dramatic process; as watching several of the television programmes associated with Units 11–15 should convince you. The mechanisms underlying morphogenesis are particularly fascinating—'how are shapes generated?' As you have already seen, growth can be a morphogenetic agent but other mechanisms are probably much more important in general. Two of these, *cell movement* and *pattern formation* (the arrangement of cells to form patterns or arrays), are dealt with in detail in Unit 14.

morphogenesis*

cell movement* **pattern formation**

Though, for convenience, we have subdivided development into three separate component processes, they are often, indeed generally, interdependent. For example, growth *can* be a morphogenetic agent; likewise our *recognition* of patterns of cells (a part of morphogenesis) depends on *cell differentiation* making those patterns visible, though as you will see underlying 'invisible' patterns are very important to morphogenesis (Unit 14). So when we deal with each component process separately, always bear in mind its likely dependence and influence on the other component processes.

4.3 Development in multicellular plants

Plants show many variations on the pattern that is fairly standard in animals. In the mosses, for example, the haploid products of meiosis may form, by mitotic activity, large multicellular organisms (called gametophytes because they produce gametes). The history of the diploid zygote does not necessarily involve a period in which the whole organism is immature followed by full maturity. Plants maintain regions of growth, differentiation and morphogenesis throughout life. Much of a mature tree is dead, but much is undergoing development. It is also a major difference between plants and animals that plant cells are not free to move relative to each other. Change in form in a plant is therefore always linked with growth. Plants offer admirable material for the study of many problems in developmental biology, notably the origin of polarity in cells and the control of differentiation.

4.4 Development in unicellular and other organisms

Unicellular plants and animals may be highly differentiated. Some of the unicellular algae are very large and lend themselves to the kind of experiments that are difficult on other material. Amoebae have rather little by way of constant form or constant cytoplasmic differentiation, but have been important in nuclear-transfer experiments (Section 8.3). The cellular slime moulds are remarkable in spending much of their life-cycle as single cells. Under certain conditions neighbouring cells may aggregate to form a differentiated multicellular body. They are therefore of particular interest in studies on intercellular communication and on morphogenesis (see the television programme, *What is development?*). Studying these simpler systems can often give clues as to mechanisms involved in development in higher multicellular organisms: they therefore constitute *model systems*. You will meet several useful unicellular model systems in this Course as a whole. In the TV programme, *What is development?* we consider the particular use of model systems in studying developmental biology.

model system*

Summary of Section 4

1 There are several episodes in the development of all organisms (at least virtually all multicellular animals) which are common.

2 These episodes occur in a particular temporal sequence.

3 Virtually all of development can be viewed as consisting of three component processes; growth, cell differentiation, morphogenesis.

Objectives and SAQs for Sections 3 and 4

Now that you have completed these Sections, you should be able to:

★ evaluate evidence attributing a role to genetic information (genotype) in development.

★ evaluate evidence attributing a role to environmental factors in development.

★ give appropriate examples, or evaluate given examples, of areas of everyday life where research in developmental biology has particular relevance.

★ given the names of certain developmental episodes occurring in animals, place them in correct temporal sequence.

★ identify or briefly define three central component processes common to the development of many organisms.

★ describe or identify from given data those features that are characteristic of (a) cleavage; (b) gastrulation.

To test your understanding of these Sections, try the following SAQs.

SAQ 3 (*Objectives 3 and 4*) Thalidomide administered to pregnant rats causes little fetal damage. This could be because:

(i) rat cells are less susceptible,

(ii) the rat placenta does not pass thalidomide to the fetus as readily as does the placenta of monkeys, men or rabbits,

(iii) rats excrete thalidomide exceptionally rapidly,

(iv) rats effectively metabolize thalidomide into harmless substances.

Suggest how one could test each of these possibilities experimentally.

SAQ 4 (*Objectives 1, 2, 4, 6 and 8*) Which of the following statements are *true* and which are *false*?

(a) There is no genetic basis at all to ageing.

(b) In the formation of all blastulae from fertilized eggs there is an increase in size due to an increase in cell number.

(c) Cytodifferentiation occurs in multicellular and some unicellular organisms.

(d) Gastrulae in higher animals contain three cell layers.

(e) In most species spermatozoa are larger than ova.

(f) Cell division cannot occur without concomitant growth.

SAQ 5 (*Objectives 1 and 5*) Place the following developmental episodes in their correct temporal sequence:

fertilization, gastrulation, oogenesis, cleavage, spermatogenesis.

5 Methods of study

In this Section we consider some of the basic techniques by which development is studied; this is discussed further in the television programme, What is development?

What, then, are the methods by which development can be studied?

5.1 The descriptive method

There are limits to the rewards of simply watching intact embryos developing, though in recent years a number of important ideas have come from doing just that or from, what is almost the same, analysis of time-lapse films of developing systems. For example, the suggestion that gastrulation in sea-urchins is accomplished partly by mesodermal cells throwing out long processes that first anchor to the roof of the blastocoel and then contract to pull in the mesoderm, has been supported by cine-records of the invagination process (Figure 6).

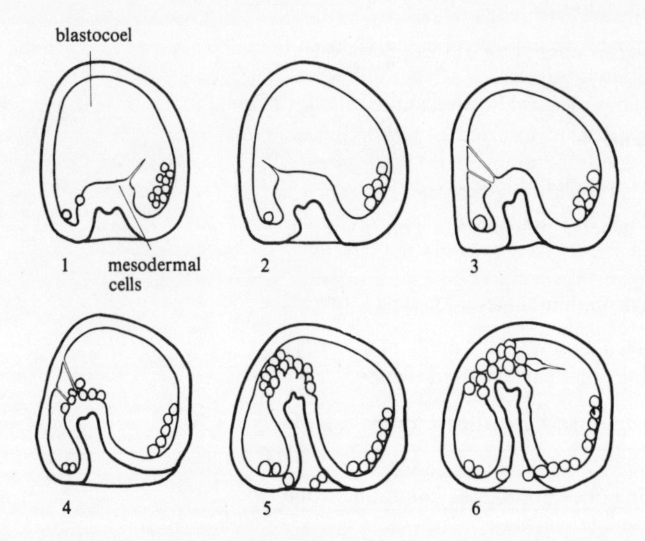

FIGURE 6 How the long processes of the mesoderm cells help invagination. 1–6 indicate consecutive stages in this process.

Descriptive study may have to overcome natural barriers: for example, in viviparous animals (i.e. bearing live young) normal developmental processes cannot always be watched. Nowadays, however, mammalian eggs can be cultured *in vitro* for some time, during which they develop apparently normally. Progress in the techniques of culturing eggs and embryos means that over half the period between conception and birth in rats can be studied *in vitro*. (*In vitro* means literally 'in glass' and in this context means outside the intact animal. A cell or molecule inside an intact animal is *in vivo*.)

Description can be important in answering questions such as: are different rates of cell division in different parts of a developing system responsible for the observed changes in shape that it undergoes? Counts of cells in mitotic division, or estimates of the rate of synthesis of DNA in the nuclei assist in finding the answer.

Description at the molecular level is also a necessary preliminary to understanding. Red blood cells are full of haemoglobin. When does it first appear in the cells that are their progenitors? When do neurons first begin to transmit impulses? The answers to such questions may come from technically sophisticated investigation, but they are still only describing what happens, in the same sense as does the answer to the question, 'when does the arm first appear?'

5.2 The manipulative approach

The methods of analysis, at the level of both cell and whole organism, are several. We can test the behaviour of a part of the embryo when it is isolated from the rest. The methods of *tissue culture* allow cells to be cultured outside the body for long periods (Figure 7). For example, will cells whose normal fate is to become neurons do so if they are deprived of contact and association with the rest of the embryo?

tissue culture*

Figure 7 provides an answer. A great deal has been learned from cell culture about the extent to which developmental processes are carried out in accordance with interactions between cells.

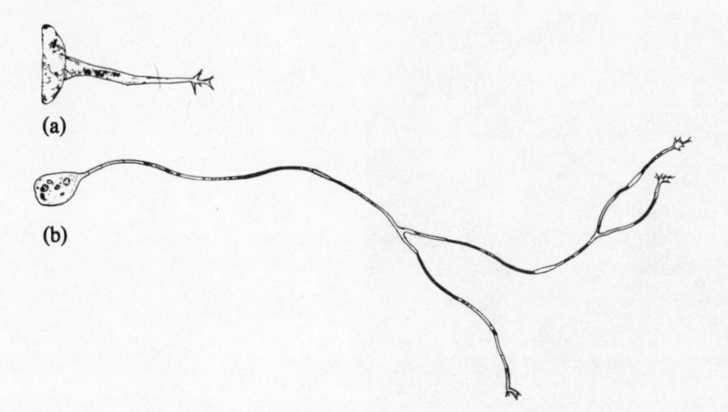

FIGURE 7 (a) A nerve fibre beginning to grow out from a cell forming part of a small mass. (b) The same fibre a few days later, after it has become branched and greatly elongated.

cell/tissue grafting*

A cell (or group of cells) or tissue can also be transferred or *grafted* from one part of an embryo to another. Are their developmental fates altered by this? If so, what happens to graft and host if they are of different species? Such methods were first used at the intercellular level of analysis. They can also be used at the cellular level. An egg fertilized by sperm of a foreign species has received a special kind of graft between two species. If its own nuclear material is removed or killed, we can produce an egg whose cytoplasm comes (mostly) from one species, but whose nucleus comes wholly from another. How would such an egg develop? In recent years a new method (*nuclear transfer*) has been widely used to study the development of eggs robbed of their own nucleus, but given one from a tissue, or a tumour cell, of an older member of their own species[9]. Even cytoplasm can be grafted. Most of the yolk can be removed from a fowl's egg and replaced by yolk from another fowl's egg. The relatively tough outer layer, or cortex, of the amphibian egg has been grafted onto other eggs of the same species in place of parts of their own cortex.

nuclear transfer*

Such experiments help to localize the areas or regions of cell or embryo that are responsible for particular aspects of the behaviour of the whole.

Less easy to analyse are the actual biochemical pathways of importance in development. It is, indeed easy to show that killing cells stops them developing! What is more difficult is to block a specific synthetic activity and leave the cell alive, but with a restricted or altered capacity to develop. At the present time it is possible to treat cells with drugs whose major effect is to inhibit the synthesis of a whole class of macromolecules—DNA, RNA, or protein, for example. Even so, the drug is unlikely to be absolutely specific. It is also possible to interfere with the metabolism of many of the smaller molecules, which are precursors of the biological macromolecules or are in other ways involved in their synthesis. In Units 12 and 13 you will see something of the molecular changes underlying development.

There are of course other techniques (see the television programme, *What is development?*) but now that we have briefly discussed some of the overall processes and techniques of developmental biology, it is convenient to consider certain points in a little more detail.

6 What changes, and what remains constant, during development?

The early researchers of development were rightly impressed by the dramatic visible changes in gross size, shape and structure that accompany development. Later, it became clear that at the level of the cell there were equally dramatic changes, and later still the physiology and biochemistry of embryos were shown to undergo fundamental change, too. Before we ask if anything remains constant during development (Section 6.4) it may be well to consider examples of things that change (Sections 6.1–6.3).

6.1 Form and structure

The shape of adults and their gross anatomy may be very complex, but in eggs these features are simple (Figure 8).

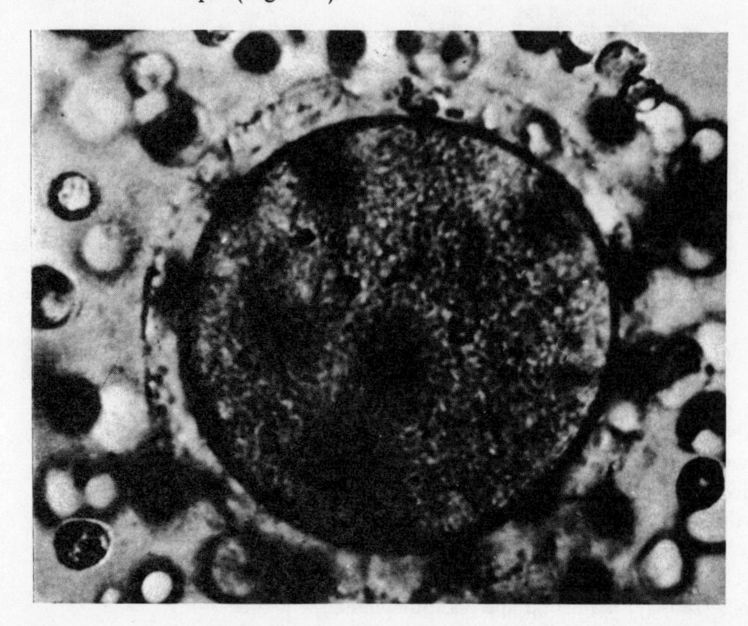

FIGURE 8 A photomicrograph of a human ovum. (By courtesy of W. J. Hamilton.)

The first important generalization that emerges from the study of developmental anatomy is that eggs do not become transformed into adults by the most direct conceivable path. A simple example will make this point. Neither a human egg nor an adult human has a free tail. Yet a human embryo does have one for a short while (Figure 9).

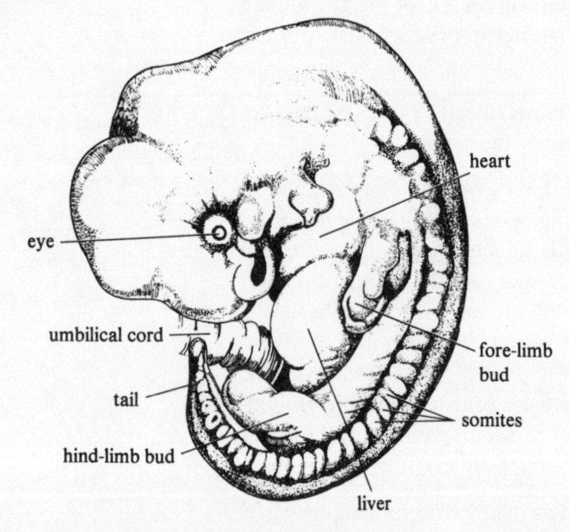

FIGURE 9 A 30-day human embryo.

Similarly, such terrestrial animals as birds do not, as adults, have gill slits, but their embryos for a short period do. These are both examples of developmental conservatism, as human ancestors had tails and avian ancestors were fish (Unit 3). During the development of their ancestors, tails and gills had to be formed and the descendant follows the ancestral path of development more accurately than would be expected if development were as direct as possible*. However, it is

* It follows that a comparison of embryos sometimes suggests an evolutionary relationship that is not at all evident from the comparison of their adults. The most famous such case is that of the echinoderms, the acorn worms, and the chordate animals. Their embryonic anatomy, especially in respect of the coelomic cavities, has similarities which invite a belief in a common ancestry. Modern views support the 'laws' put forward by von Baer (1792–1876), the founder of comparative embryology. They were based only on anatomical findings and run, in parapharase, as follows:

1 In development general characters appear before specialized ones.

2 Specialized characters develop from general ones.

3 During the development of an animal, there is a progressive departure from the form of other animals.

4 The early stages of higher animals resemble the early stages, but not the adults, of lower animals.

important to realize that embryos, like adults, have to cope with their environment, and that developmental conservatism does not prevent the evolution of striking adaptations to the transient needs of embryonic life. Thus birds and reptiles have evolved a number of structures that serve only to facilitate embryonic nutrition, respiration and excretion within an egg shell that cannot be too porous, or water loss from the embryo would be lethal. Some of these structures are discarded at hatching.

The higher placental mammals are descended from reptiles and have so modified the structures in question that they serve the special needs of life *in utero*. Here there is no problem of water loss, and nutrients can be drawn from the maternal blood and excretory products returned to it via the placenta. The shell has disappeared and structures associated with the absorption of yolk (the yolk-sac) or with respiration and the storage of nitrogenous waste (the allantois) in reptiles are used to establish a close neighbourhood between maternal and fetal blood circulation (Figure 10).

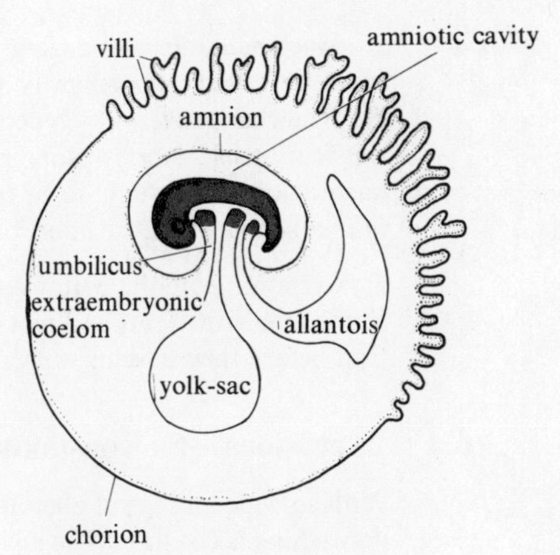

FIGURE 10 An embryo (pink) of a placental mammal, surrounded by embryonic membranes.

6.2 Function

Before cell differentiation and morphogenesis are accomplished, many physiological activities, normal to the adult, are impossible. Blood flow cannot easily precede the formation of vessels and of the heart. Neural control of the heart cannot precede the establishment of its nerve supply. But the development of function does not always take the form of a jump direct to the adult mode. As an example, we may take the excretion of nitrogenous by-products of metabolism. These will be formed from the breakdown of, for example, amino acids, purines and pyrimidines. Ammonia is in some ways a good candidate for excretion of nitrogen. However, its toxicity at any but the lowest concentrations makes it impracticable unless an exceedingly dilute, and hence copious, urine can discharge it as soon as it is formed. This is possible for animals living in freshwater and tadpoles do, in practice, excrete ammonia (Units 22 and 23). Shell-bound embryos such as those of birds would soon be poisoned by ammonia; nevertheless they do produce it temporarily, then proceed to produce urea which is less toxic but very soluble, and finally to uric acid which is sequestered as a harmless, fairly insoluble sludge to be discarded at hatching (Figure 11).

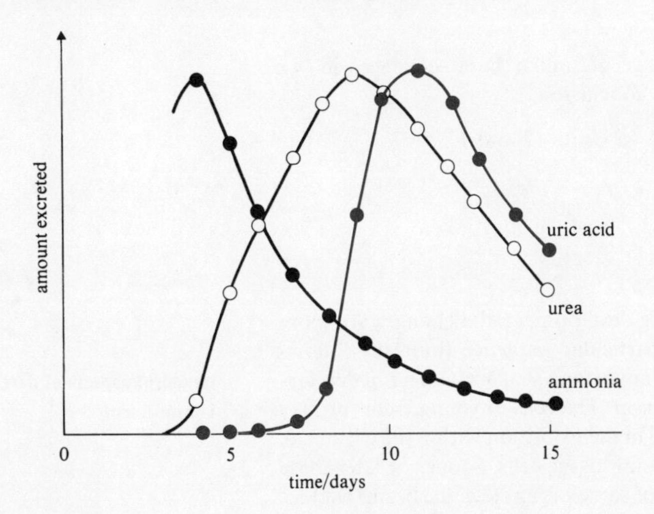

FIGURE 11 The excretion of ammonia, urea and uric acid during development in a bird's egg.

6.3 Chemistry

It is impossible to be sure that eggs do not contain minute amounts of substances that are present in great quantities in adult tissues. It is also important to appreciate that an egg or young embryo may contain materials that were synthesized in the mother's body and are passively inherited. Nevertheless, it is quite clear that eggs and embryos differ from adults in overall chemical composition, in synthetic activity and in some metabolic pathways. A few examples will make this apparent.

1 An embryo (or fetus or new-born mammal) may contain antibodies that it has not itself synthesized but has passively inherited from its mother. These antibodies offer some protection against pathogens until such time as the young animal can react effectively to antigens by making its own antibodies. Yolk reserves are a less exotic example of passive inheritance.

2 Mammalian fetuses during most of fetal life have a haemoglobin in their red blood cells that differs slightly in detailed structure from the haemoglobin of adult life. The changeover, which depends on the synthesis of the adult haemoglobin by the fetus itself, starts before birth and in rare cases may fail to occur. Fetal haemoglobin is thought to be better adapted to the problems of oxygen transport *in utero* than would be adult haemoglobin.

3 Certain metabolic pathways develop fairly late in development. For example, in mammals the fetus and new-born rely on glycolysis for carbohydrate metabolism before they develop a TCA cycle.

6.4 Exceptions—the constants

With so wide a range of phenomena changing during development it is fair to ask if anything is constant. The answer is that two important features do not change. First, the egg, though a very unusual cell, is a cell in all essential features and its descendants remain cells in the same sense. Thus the pattern of cellular organization and the major components of cells—nuclei, mitochondria, ribosomes, membrane systems—characterize both the egg and nearly all types of cell derived from it. There are some exceptions; for example, post-natal mammalian red blood cells have no nucleus, and striated muscle cells are multinucleate. The second invariant is chemical and refers to the DNA of the chromosomes. The structure of DNA simultaneously permits it to carry specific information in the sequence of bases of which it is largely composed and permits this information to be replicated where one DNA molecule provides the template for the production of two 'daughter' molecules.[10]

The evidence for the identity, or near identity, of the DNA of the zygote's nucleus and of the nuclei of the adult cells descended from it is of several kinds.

1 Generally, the visible behaviour of chromosomes during mitosis is consistent with precise replication.

2 Nuclei from adult cells can, in some instances, be used to substitute for the nucleus of a fertilized egg.[11]

3 Many of the proteins that are specified by the DNA are the same in different tissues and in an animal of any age.

4 Direct methods of testing the percentage of similar base sequences in two nucleic acid preparations lead to a similar conclusion.

We shall elaborate on the identity of DNA in Units 12 and 13.

7 The process of commitment

Though some things remain constant during development, the changes are more dramatic. Furthermore, they occur in a particular sequence (butterfly follows pupa follows larva). We can illustrate this important *temporal aspect of development* by reference to the *process of commitment*. The cells of young embryos may differ from each other in size and shape and in their position within the organism. They do not possess the characteristics of adult tissue cells. Sooner or later these must be acquired. Study of the cell-lineage of some organs (e.g. the brain) enables one to say at a very early stage in development that from a particular, localized population of cells, and from no others, will that organ be formed. At the time when such a statement can first be made the cells in question may, individually, be indistinguishable from others in appearance. What is most likely to characterize them is their position in the embryo. How far are appearances deceptive? Are such cells really committed to their normal fate? The answer varies with species and with the stage of development.

temporal aspects of development*
commitment*

In some species blastomeres at the two-cell stage show differences that are associated with the fate of their descendants and cannot be reversed. In others, the blastomeres have entered into quite firm commitments by the 8-cell or 32-cell stage. That is, if they are maintained in any situation compatible with survival and cytodifferentiation, they will always follow the same path of differentiation as they would do in the intact embryo. They are said to be *determined*. In other words, we must distinguish between at least two degrees of commitment. A particular cell, or group of cells, that can be identified in a young embryo may always, in normal development, follow a certain fate: such cells are said to be committed. But if abnormal conditions can change their fate then final *determination* has not taken place.

determination*

In other species, including vertebrate animals, the process of determination is slower, takes place in some regions of the embryo before others, and is progressive. Cells are committed in steps, each step restricting the future possibilities open to them until finally their normal fate is determined. Such progressive determination made the analysis of differentiation easier to study and, in particular allowed one kind of intercellular interaction—*embryonic induction*—to be investigated.

embryonic induction*

7.1 Embryonic induction

Embryonic induction occurs when a cell population that still has some options open to it (e.g. to become epidermal or nervous tissue) has the decision made for it by its contact with another cell population—the so-called *embryonic inductor* or *inducer*. Embryonic inducing systems are responsible for the choice between epidermis and the crystalline lens of the eye and between opaque skin and transparent cornea, as well as being necessary for the formation of the central nervous system as a whole. They are involved in details of the formation of lungs, kidneys, gonads and many other organs. One possible value of inductive relations is that they can ensure a good fit between different components of a compound organ. The lens of the eye must be in the right position relative to the retina, to work well optically. Its induction by the future retinal cells is therefore apposite.

embryonic inductor*

Embryonic induction shows dramatically that the path of cytodifferentiation adopted by a cell population can depend upon its immediate environment, the adjacent cells. Is the stimulus provided by the inductor specific? Does it provide detailed instructions to the induced cells? The answer to the first question is a qualified yes. Although a wide range of non-specific stimuli (e.g. damage, heat, centrifugation) can mimic some of the effects of some inductors, it appears that specific chemical agents are involved in the best investigated cases. The answer to the second question is no. The inductor is a trigger that sets off a chain of processes in the target cells that must therefore themselves contain the bulk of the information needed to become differentiated.

It is characteristic of those inductors that are easiest to test that they do not appear to be specific to a species. So if a piece of gastrula from the dorsal lip region is grafted from an amphibian of one species into a young gastrula of another species of amphibian a secondary embryo is thus induced (Figure 12).

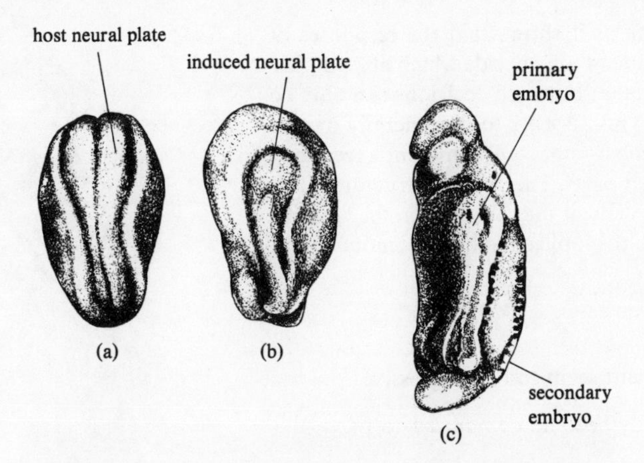

FIGURE 12 Species non-specificity of inductors. A piece of gastrula from the so-called *dorsal lip* region was grafted from an amphibian of one species into a young gastrula of another species of amphibian. As can be seen, a secondary embryo is thus induced. (a) and (b) are different views of the host and induced neural plate regions; (c) is a side-view of the embryo at a later stage.

So, as with many vertebrate hormones, embryonic inductors will exercise their effects on animals of widely different genotype. The difference between the brain of a human and a mouse may owe little or nothing to the chemistry of the stimuli that commit the cells to form brain tissue in the first place, but may owe most or all to the response of the ectodermal cells to the stimulus (see Figure 13.)

See the *S202 Picture Book* for details of *Triturus*.

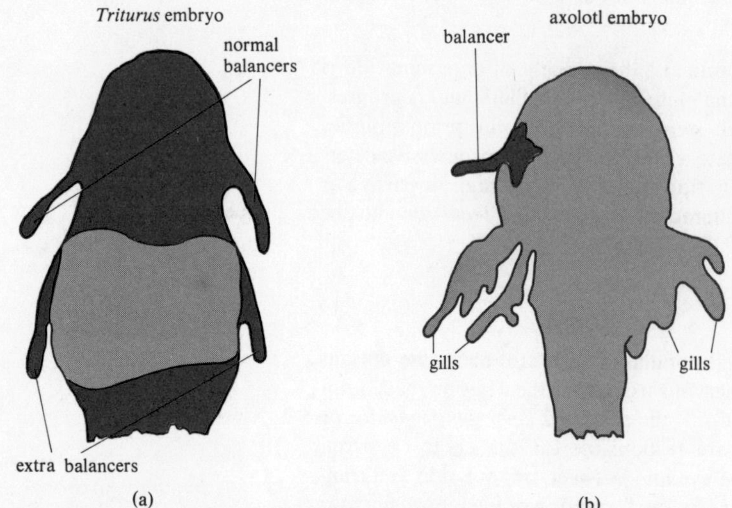

FIGURE 13 *Triturus taeniatus* (newt) normally develops an organ called the *balancer*; another amphibian, the axolotl, does not. (a) A piece of axolotl gut was grafted into an embryo of *Triturus* which has as a result developed extra balancers. (b) A piece of trunk epidermis of *Triturus* grafted onto the head of an axolotl embryo has given rise to a balancer (*left side of head*); no balancer is formed from the normal axolotl epidermis (*right side*). The axolotl therefore possesses the required stimulus for balancer formation (a) but its epidermis fails to react to it (b).

competence*

Thus both 'signal' (inductor) and *competent* tissue are needed to elicit a response. This point is made if we look at another system, the head of amphibians. If at an early stage, tissue that is normally destined to become belly epidermal ectoderm is transplanted to the head, it will behave there as would normal head ectoderm and form typical local structures. Such behaviour, in which cells adapt to unpredictable change to give a normal end-result, is known as *regulation.* (An extreme case is seen in monozygotic twinning where one embryo gives rise to two normally formed ones.)

regulation*

However, as inductive stimuli can work across genetic boundaries, we may ask whether belly ectoderm from one species transferred to the head of another forms structures that are specific to the host or to the donor. A classical experiment gave the answer: the belly ectoderm of a frog transplanted to the head of a newt gave its host typical head tissues—but they were typically frog (Figure 14).

This result is of great theoretical importance as it shows that the responses of developing cells to stimuli from their environment may decide which of a number of possible paths of cell differentiation they take; but such decisions can only be within the limits of their genetic repertoire. This appears to be generally true of cell differentiation. For morphogenesis the most important apparent exceptions concern quantitative features of organs or their parts. The size of a structure is not always typical of the species contributing its cells; it may depend on the 'magnitude' of the inductor signal. We shall return to the topic of embryonic induction in Unit 14.

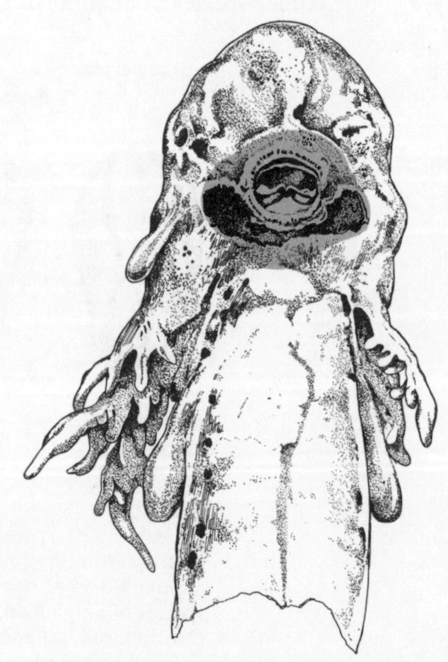

FIGURE 14 Larva of a newt in which the ectoderm in the mouth region was replaced by ectoderm of a frog embryo. The grafted ectoderm shown as pink has developed horny jaws and teeth typical of frog tadpoles.

Summary of Section 7

1 The changes that occur during development seem to be progressive.

2 Different cells or groups of cells gradually become committed to particular fates.

3 Committed cells can have their fates altered, for instance by experimental manipulation. Ultimately, though, the commitment becomes irreversible, the cell (or group of cells) is determined.

4 Commitment and determination depend on interactions between neighbouring cells at the 'right' time, as demonstrated by the phenomena of embryonic induction and competence.

5 Embryonic induction involves signals (embryonic inductors) triggering complex chains of responses in the tissue that is induced.

8 Nucleo–cytoplasmic relations

The phenomenon of embryonic induction demonstrates the important influence of interactions between cells during development. But what of the subcellular changes themselves? These involve interactions between the genetic information in the form of DNA in the nucleus and chemical 'signals' emanating from the cytoplasm.

In order to study the interactions of nucleus and cytoplasm, and their relevance to the later history of a cell and its progeny, a number of techniques can be used. Two useful techniques involve *reciprocal hybrids* and *species hybrids*.

reciprocal hybrids*
species hybrids*

8.1 Reciprocal hybrids

If a cross between two genetically distinct populations A and B is made both ways, that is, female A × male B and male A × female B, the two kinds of progeny should be indistinguishable in all respects in which nuclear information only is concerned. In practice, reciprocal hybrids may show some maternal influence (e.g. in the rate of cell division during very early development), but usually do not. If they do differ, the difference may be due to a maternal effect attributable either to the preponderantly maternally derived cytoplasm of the fertilized egg or (in the case of mammals, for example) to a maternal environment in post-fertilization life. Thus C57B1 mice (a highly inbred strain) normally have six vertebrae in the lumbar region of the spine while C3H mice (another strain) have only five. The reciprocal hybrids follow the mother, but the transfer of fertilized ova from the mother to the uterus of another mouse has shown that the effect is due to the uterine environment.

8.2 Species hybrids

Species hybrids, that is hybrids derived from matings between two different species, have certain advantages for work on nucleo–cytoplasmic interactions. It is quite common for small differences between closely related species to be apparent quite early in development. For the most part intra-specific genetic differences are detectable only at late stages of development. Most attempts to hybridize between species fail, but some succeed. Several levels of success can be recognized. It is rare for species hybrids to be fully viable and fertile, though cases are known among wild ducks, cyprinid fishes and flowering plants. More often, if viable, they are sterile (as with mules and hinnies) though they may show what is known as hybrid vigour, whereby the hybrid is healthier than either parent. More often still they die young (sheep–goat hybrids). Many inter-specific crosses in amphibians and echinoderms die during gastrulation—a time at which transcription of the DNA begins in earnest as other evidence shows. Some other attempts at inter-species hybrids appear to succeed but success is illusory. Sometimes the sperm activates the egg but its nucleus does not participate in development, which is therefore parthenogenetic. Sometimes, however, it is possible to remove or destroy the egg's nucleus while successfully fertilizing it with sperm of a foreign species. Any subsequent development will be of an organism the vast majority of whose cytoplasm was originally of one species, but whose nucleus comes from another. Such creatures have been reared through early development in crosses in echinoderms, in amphibians and in silkworms. The species used in such crosses are fairly closely related and so they differ very little in early embryogenesis. Nevertheless, it appears that the nucleo–cytoplasmic hybrid follows its maternal (cytoplasmic) species pattern in such matters as the rate of cell division during cleavage and gastrulation; but once cell differentiation begins, it is the paternal (nuclear) species that determines the nature of the embryo.

One celebrated case gave a different result. A nucleo–cytoplasmic hybrid between two newt species was known to die young. By transplanting cells from this hybrid before it died it was found that they could survive well on a normal host of a third species. At metamorphosis (i.e. the change from larval form to adult), weeks after the graft had been made, the grafted cells showed tissue properties characteristic of the species that had provided egg cytoplasm. This result invites further study, but reminds us that cytoplasm may have sources of information of its own.

8.3 Nuclear equivalence

Hybridization, referred to in the last Section, only allows us to study known or suspected genetic differences between different kinds of organism of the same or different species. But part of our problem is to discover whether or not different cells of the same organism have acquired, during development, stable or irreversible differences that might serve to explain the differentiated state. In other words, are the nuclei of differentiated cells all equivalent in terms of their genetic information or not? This can be tested both indirectly and directly.

If the process of differentiation necessarily involved a specific, permanent, nuclear change, then a differentiated cell could not change its differentiation. Neurons could not become muscle cells, for example. Yet changes in tissue type, called *metaplasia*, have been reported in many organisms. The most convincing is probably the conversion of differentiated iris margin cells into lens cells in newts. Here, in a process known, after its discoverer, as 'Wolffian regeneration', surgical removal of the crystalline lens of the eye is followed by its replacement. The cells that form the new lens are, however, originally quite unlike lens cells—they are, among other things, pigmented. Yet, though this shows that iris margin cells in the course of their differentiation have not lost the nuclear information necessary for lens differentiation, it might be a special case. As you will see further from Units 12 and 13, plants provide more telling evidence in that a whole, fertile, adult plant (e.g. carrot, celery) can be obtained from the culture of a single isolated tissue cell. The experiments of Briggs and King, and Gurdon, also led to a more direct approach. They were able to remove from a frog's egg the whole of its own nuclear apparatus. Such an enucleated egg normally dies very quickly. But after introducing into it a cell nucleus taken from a somatic cell of another embryo, the resulting egg with a 'foreign' nucleus may survive and develop.[12] It follows that the donor nucleus has been able to provide everything necessary for development that the nucleus of a normal zygote would. (The results of these experiments are discussed in more detail in Units 12 and 13.) The conclusions force us to look again at the organization of the cytoplasm of the egg cell. It is here that we must seek the origins of the different environments that call forth from the nuclei different parts of their common repertoire (Unit 15). For now, it suffices to say that interactions between nucleus and cytoplasm seem to be vital to development.

metaplasia

Objectives and SAQ for Sections 5 to 8

Now that you have completed these Sections, you should be able to:

★ name or recognize or briefly describe or analyse those things that change during development and those that remain constant.

★ name and briefly describe or recognize the use of techniques or methods of study relevant to developmental biology.

To test your understanding of these Sections, try the following SAQ.

SAQ 6 (*Objectives 1, 7 and 9*) Which of the following statements are *true* and which are *false*?

(a) Embryonic inductors are not generally species-specific.

(b) Parthenogenesis proves that the sole role of a spermatozoon is to activate the ovum to develop.

(c) A cell is said to be committed only when, no matter what changes are made in its environment, it develops in isolation as it would in a normal intact embryo.

(d) Embryonic induction is important in ensuring the correct juxtaposition of different tissues.

(e) The experiments of Briggs and King, and Gurdon, show that cell-differentiation does not involve a loss of genetic information from the cell nucleus soon after cleavage.

(f) The shape of an adult organism is generated by morphogenesis involving the most direct transformations from the egg.

(g) One of the main purposes of a model system is to simplify the process one is interested in.

9 The internal organization of the egg

The usual starting point for development is considered to be the egg. Although eggs are anatomically simple they always show some structure. Those that are spherical in overall shape are nevertheless polarized in their internal arrangements. The nucleus may lie eccentrically and the point on the surface nearest to it is then known as the *animal pole* (the opposite pole is known as *vegetal*, Figure 15). Generally, eggs and embryos are shown oriented with the animal pole uppermost.

animal pole*
vegetal pole*

The distribution of cytoplasmic constituents is polarized, too. Yolk, pigment, mitochondria, may all be arranged in a defined and constant way in relation to the *animal–vegetal axis.* In some species the cytoplasm may in addition possess visible localized concentrations of pigments.

animal–vegetal axis*

It is clear that such cytoplasmic localizations, if they persist during cleavage, must mean that the nuclei of the young embryo, although themselves equivalent in genetic composition, find themselves in different cytoplasmic environments. These differences *may be* crucial to subsequent development (Units 12 and 13, and 15). Although the number of different environments may at first be small, once any diversity in cell properties is established it can be used to generate more by cellular interactions. Thus the changes 'snowball'.

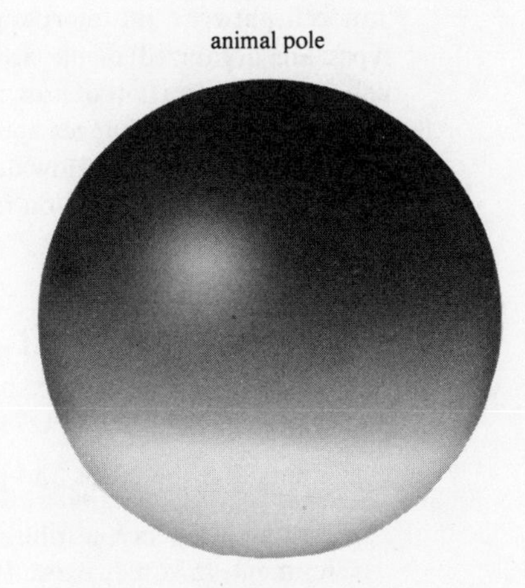

FIGURE 15 Diagram of a frog's egg, showing animal and vegetal poles.

There is direct experimental evidence that the cytoplasmic localizations of the egg do determine the fates of the cells to which they are allocated. For example, disturbing the distribution of egg materials by mechanical means (usually centrifugation) may cause corresponding disturbances in cell differentiation. Removal of parts of the egg's cytoplasm can lead to a later absence of the corresponding embryonic tissue. All this can only mean that the inheritance that an individual receives from its parents is in four forms:

1 Coded, invariant, genetic information in nucleic acid.

2 Cellular organization of a general kind.

3 Specific morphogenetic information in the form of a non-uniform egg cytoplasm that will provide the nuclei of the embryo with different environments.

4 Passive materials of functional, but not necessarily morphogenetic, significance (yolk, antibodies, etc.).

It is important to appreciate that although the developmental information stored in the egg cytoplasm is created in some instances during its life in the ovary, it may be controlled by the maternal genotype, There are several instances known of the Mendelian inheritance of egg–cytoplasm properties. In these, the genotype of the mother, not that of the fertilized egg, determines some feature of development.

Yet, although it is certain that we shall discover more and more cases of the effects of maternal genes upon egg structure, there is also reason to suppose that some of the information contained in the egg is not genetically determined in this fashion. For example, in many, but not all, insects the elongated eggs lie in the ovary with their long axes parallel to the long axis of the mother's body. Furthermore, in its subsequent development this ovarian *polarity* is retained by an embryo—the forward-pointing end of the egg in the ovary becoming the anterior end of the embryo. Thus, if we consider the egg as providing the starting point for development we find that the egg itself depends on its history in the mother for its subsequent activity. We seem to have recreated the 'chicken or egg' paradox!

polarity (egg/embryo)*

10 Summary and conclusions

In this Unit we have tried to present a broad view of developmental biology to show how genotype and environment interact to produce an adult organism from a bud, seed or egg. This has meant the introduction of a large number of systems, ideas and terms which may well be new to you. It is not our intention that in this Unit alone you should become familiar with all these concepts, but you should by now have some understanding of the different factors that go towards development, and how they are coordinated. Though they are interdependent, it is necessary, in order to study and discuss them, to consider these factors in isolation from each other, at least to some degree. In this Unit we considered three central processes that occur in going from egg to adult—*growth, cell differentiation* and *morphogenesis*—and how they are influenced by the genotype and the environment. In the rest of the Development Block, the same strategy is adopted but some points are considered in much more detail. Units 12 and 13 deal with what is currently known about the subcellular changes that underlie cell differentiation, and, in particular, with how such changes are controlled. In Unit 14 we turn to how cell patterns and morphogenesis are controlled, that is, how different cell types are organized in the arrays characteristic of tissues and organs. This involves a consideration of how cells move, how they interact with each other, and to what extent the changes are genetically programmed. Finally, in Unit 15, we turn full circle and ask 'How does it all begin?' What is it, if anything, about the egg that sets development on its particular path?

Objectives for Unit 11

At the end of this Unit you should be able to:

1 Understand the terms and principles marked with an asterisk in Table A.

2 Evaluate evidence attributing a role to genetic information (genotype) in development. (*SAQs 1, 2 and 4*)

3 Evaluate evidence attributing a role to environmental factors in development. (*SAQ 3*)

4 Give appropriate examples, or evaluate given examples, of areas of everyday life where research in developmental biology has particular relevance. (*SAQs 3 and 4*)

5 Given the names of certain developmental episodes occurring in animals, place them in correct temporal sequence. (*SAQ 5*)

6 Identify or briefly define three central component processes common to the development of many organisms. (*SAQ 4*)

7 Name or recognize or briefly describe or analyse those things that change during development and those that remain constant. (*SAQ 6*)

8 Describe or identify from given data those features that are characteristic of (a) cleavage; (b) gastrulation. (*SAQ 4*)

9 Name and briefly describe or recognize the use of techniques or methods of study relevant to developmental biology. (*SAQ 6*)

ITQ answers and comments

ITQ 1 The earlier the somatic mutation occurs in development the greater its likely effect, as the altered cell will still be destined to go through many rounds of cell division and hence produce a larger number of mutated cells. A somatic mutation later in development, when cell division is almost over, would give rise to an altered cell which would then produce only a few descendant cells and hence be less likely to produce any great effect.

ITQ 2 The explanation is that:

(a) low doses cause little damage to the sperm nucleus, so development is normal;

(b) higher doses damage some or all of the chromosomes in such a way that development is affected;

(c) still higher doses damage the chromosomes so severely that they cannot participate at all in development which is thus equivalent to parthenogenesis;

(d) at the highest doses not only are the chromosomes damaged but the whole sperm is killed and unable to fertilize (or *activate*) the eggs.

SAQ answers and comments

SAQ 1 As the resultant zygote will contain sex chromosomes X and Y the child will be male.

SAQ 2 If red colour is dominant to white, it is possible that each parent plant is heterozygous for the red colour; that is, it contains one red-producing allele and one white-producing allele of the same gene (*R*//*r* where *R* allele gives the red phenotype, *r* white). Such plants can each give rise to the same two possible types of gamete containing *R* or *r* alleles. Therefore the possible types of zygote are:

R//*r* red
R//*R* red
r//*r* white

SAQ 3 (i) and (ii) would be excluded if thalidomide were administered directly to the fetus, by-passing the placenta, and the fetus subsequently showed specific deformities. Short-term experiments with fetuses cultured *in vitro* are also possible and longer term cultivation of fetal cells could be used to look for changes in cell properties. (iii) could be tested directly. (iv) could be tackled by administering thalidomide in which particular atoms are radioactively labelled, then looking for the label in potential metabolic products of thalidomide. However, these suggestions are not the only possible ones. For example, thalidomide itself may be harmless but humans could metabolize thalidomide to produce a toxic product while rats left the molecule intact.

SAQ 4 (a) False. Difficult to prove but different species have well-defined maximum life-spans. Longevity within a species seems 'to run in families'.

(b) False. In many species there is no increase in overall size, each cell is smaller than the original zygote.

(c) True (Sections 4.2 and 4.4).

(d) True. Ectoderm, mesoderm and endoderm.

(e) False (Section 4.1).

(f) False (Section 4.2)

SAQ 5

Spermatogenesis / Oogenesis → fertilization → cleavage → gastrulation

SAQ 6 (a) True (Section 7.1).

(b) False. Parthenogenesis can allow the development of haploid organisms containing only maternal genetic information. However, normal diploid development would also involve paternal genetic information from the spermatozoon. Thus, in diploid species the spermatozoon has at least two roles (i) to activate the ovum, and (ii) to provide genetic information.

(c) False. Such a cell is said to be *determined.*

(d) True (Section 7.1)

(e) True (Section 8.3)

(f) False (Section 6.1)

(g) False. A model system does not simplify the actual process to be studied; it *is* (one hopes) a simpler version or *model* or *mimic* of the process. By studying this simpler version the more complex process *may* be better understood (see for example the television programme, *What is development*? and Section 4.4).

References to the Foundation Course

	S101	S100
1	Unit 22, *Physiological Regulation*, Section 3.2.	Unit 18, *Cells and Organisms*, Section 18.3.1.
2	Unit 19, *Genetics and Variation*, Section 2.3.	Unit 19, *Evolution by Natural Selection*, Section 19.2.1 and 19.6.4.
3	Unit 19, Sections 2.3 and 6; Unit 25, *DNA, Chromosomes and Growth: Molecular Aspects of Genetics*, Section 7.	Unit 19, Section 19.2.3.
4	Unit 25, Section 7.2.	Unit 19, Section 19.2.3.
5	Unit 25, Sections 4.2, 4.3 and 5.	Unit 17, *The Genetic Code: Growth and Replication*, Sections 17.3 and 17.8.
6	Unit 19, Section 2.	(no equivalent)
7	Unit 19, Section 3.	(no equivalent)
8	Unit 19, Section 6; Unit 25, Sections 6.2 and 6.3.	Unit 17, Section 17.11; Unit 19, Section 19.2.3.
9	Unit 25, Section 8.	Unit 17, Section 17.13.
10	Unit 25, Sections 4.2, 4.3 and 5.	Unit 17, Sections 17.3 and 17.8.
11	Unit 25, Section 8.	Unit 17, Section 17.13.
12	Unit 25, Section 8.	Unit 17, Section 17.13.

Acknowledgements

Grateful acknowledgement is made to the following sources for permission to reproduce illustrations in this Unit:

Figure 2 J. D. Ebert and I. M. Sussex (1965) *Interacting Systems in Development*, Holt, Rinehart and Winston; *Figure 3* by courtesy of Prof. J. A. Parkes; *Figure 8* by courtesy of W. J. Hamilton.

units 12 and 13

Cellular Differentiation

Contents

(*cont.*)

TABLE A Scientific terms and principles used in Units 12 and 13

Assumed knowledge†	Introduced in an earlier Unit	Unit	Introduced or developed in this Unit	Page
adenosine triphosphate (ATP)	algae	1	*Acetabularia* experiments*	15
amino acids	allosteric inhibition	7	actinomycin D*	16
analogue	animal pole	11	Britten–Davidson model*	30
antibiotic	artefacts	4	cDNA–mRNA hybridization*	35
auxin[6]	bacteria	1	cell fusion	13
carbohydrates	blastula	1, 11	cellular differentiation*	5
cell division	cell membrane	4	chromatin*	28, 37
cell structure	cellulose walls of plants	2	competitive DNA–mRNA hybridization*	33
embryo	chromatin	4, 5		
enzymes	chromatography	5	complementary DNA (cDNA)*	35
genes[3]	cleavage	11	constitutive mutants*	23
hormone[1]	cloning	11	control of transcription*	12
insulin	degradation of proteins	8	control of translation*	12
larval stage	Diptera	2	co-repressor molecules*	26
metabolism[12]	enzymes	6, 7	cyclic AMP effect	26
metamorphosis	eukaryotes	1	DNA–DNA hybridization*	29
mitosis[4]	gastrula	1, 11	differential expression of genes*	12
morphology	glycolytic pathway	8	differential transcription of genes*	34, 42
mRNA[7]	half-life	4	Ehrlich cells	14
mutants[14]	histones	5	enzyme induction*	22
nuclear transplantation[8]	metabolic pathways	8	enzyme repression*	25
nucleic acid	nematodes	1	erythrocyte	14
nucleus	phosphoenolpyruvate	8	gastrula mDNA*	40
pancreas	prokaryotes	1	gene-activator protein	26
pH	radioactive labelling	4	HeLa cells	14
plant cell culture[5]	reciprocal hybrids	11	histone proteins*	37
polymerization	rhizoid	2	hormone–receptor complex	38
polypeptide chain[9, 13]	substrates	6	*i* gene*	24

* These terms must be thoroughly understood—see Objective 1.

† Most of these terms are explained in the Science Foundation Course. Those that are of particular importance to the understanding of this text appear with a superscript number and have a full reference at the end of the Unit.

Assumed knowledge†	Introduced in an earlier Unit	Unit	Introduced or developed in this Unit	Page
protein synthesis[9, 13]	TCA cycle	8	inducer*	22
replication[4]	vegetal pole	11	intervening sequences of DNA*	44
rRNA[11]	tissue culture	11	Jacob–Monod hypothesis*	24
spawning			long-term regulation	28
transcription[9, 13]			messenger DNA (mDNA)*	36
translation[9, 13]			multinucleate cell	13
tRNA[11]			non-histone proteins*	37
viruses[10]			nuclear transplantation*	9
zygote[2]			null mDNA*	40
			oocytes	41
			operator region*	24
			operon*	25
			plant cell culture	8
			pluteus*	41
			promoter region*	24
			protoplasts	14
			regulator gene*	23
			regulon*	26
			repressor molecule*	24
			reverse transcriptase	35
			ribonuclease*	16
			short-term regulation	28
			single copy DNA (scDNA)*	30
			stored mRNA	16, 41
			structural gene*	23
			tailoring and splicing of mRNA*	44
			temporal control	6
			tissue culture	8
			totipotency*	8
			totipotent*	5

Study guide for Units 12 and 13

In Unit 11 we outlined the questions scientists ask when considering the way organisms develop. In Units 12 and 13 we concentrate on the ways in which the variety of cellular types seen in multicullar organisms can arise from the basic cells seen in early stages of development after the formation of the zygote. We discuss what factors in the cell determine its function and how these factors are controlled to give the variety of differentiated cells.

We look at the nature of the genetic material and consider the idea that in one organism all the cells have the same genetic material. However, as you will see, the information in this material is expressed at different times during development.

The role of the cytoplasm and its effects on the activity of the genetic material in the nucleus are discussed. Then follows a Section that emphasizes that the differences in cells arise from chemical differences—specifically, differences in proteins.

We then consider the way in which bacteria could control the expression of their genetic material to produce changes in levels of enzyme proteins. This provides a model system which can be used as a basis for the detailed discussion of the control of cellular differentiation in higher organisms. And this is the subject of the final Sections of these Units.

The Units are designed to take two weeks of study. You should try to reach the end of Section 5, on the control of protein synthesis in bacteria, by the end of the first week.

There is no experiment associated with these Units but you will find that the television programme, *Differential Gene Expression*, as well as reinforcing material in the text, covers some aspects of experimental work; it is essential viewing. Note there is only one programme associated with the Units.

Units 12 and 13 rely heavily on S101*, Unit 25, Sections 4–6 and 8 (S100†, Unit 17, Sections 17.4, 17.6, 17.8–17.10, 17.12 and 17.13). You may find it helpful to spend some time revising this material. There is, though, a pre-Unit test on DNA and protein synthesis which allows you to assess your understanding of the Foundation Course material.

For Section 4, you will need some knowledge of enzymes. You studied these in Units 6 and 7, but here your knowledge from the Foundation Course should suffice.

Pre-Unit test on protein synthesis

Figure 1 shows the synthesis of a specific protein based on information in the DNA. We have divided the process into a number of stages.

(a) Label stages 1 and 2 on the diagram and explain what happens at each stage.

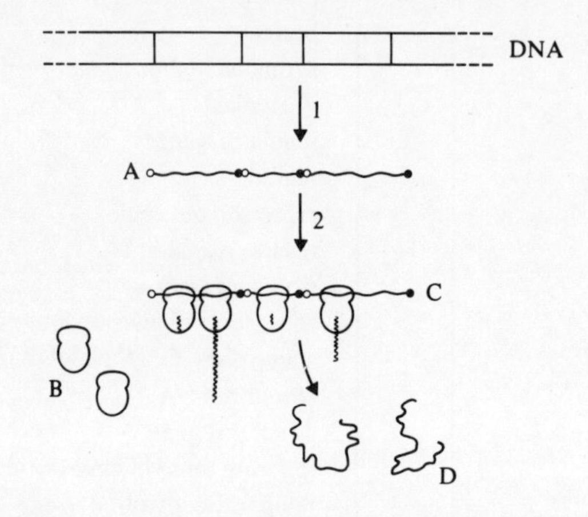

FIGURE 1 Protein synthesis.

Various chemicals or structures are needed for protein synthesis: messenger RNA (mRNA), transfer RNA (tRNA), amino acids, activating enzymes, ribosomes, nucleotides, polysomes and polypeptide chains.

(b) Which of these chemicals or structures are needed for stage 1?

(c) Which are needed for stage 2?

Some of the structures in the list are shown as A–D.

(d) Label structures A–D on the diagram.

Check your answers against those on p. 48.

1 What is cellular differentiation?

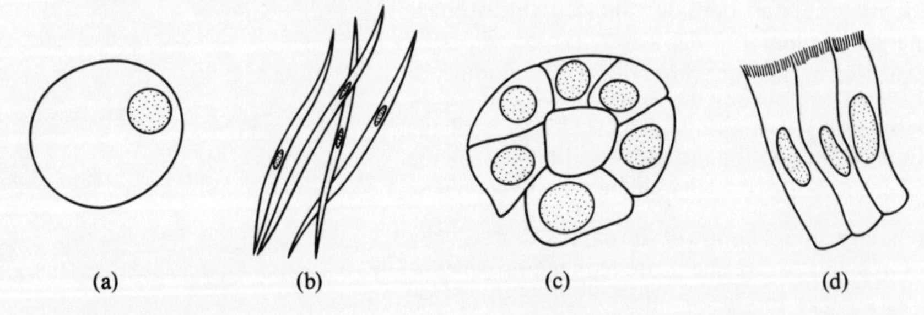

FIGURE 2 Different types of cell. (a) Unfertilized ovum. (b) Smooth muscle cells. (c) Pancreatic cells. (d) Intestinal cells—the lining of the small intestine.

Consider the cells in Figure 2. The difference in appearance between the unfertilized ovum in Figure 2a and the adult cells (Figures 2b–d) makes it clear that great changes in cell structure occur during development. Other differences are not so apparent. For example, all three adult cells have different types of protein: the muscle cells have high levels of two proteins, actin and myosin; the pancreatic

* The Open University (1979) S101 *Science: A Foundation Course*, The Open University Press.

† The Open University (1971) S100 *Science Foundation Course*, The Open University Press.

cells produce large amounts of the protein hormone[1] insulin, and the intestinal cells produce a range of enzyme proteins. These are all examples of biochemical changes in cells. Your studies of animal histology (home experiment) will also have shown you that there is a great variety of cell types in adult organisms.

During development there is a change from one basic cell type—the cell arising from the fertilized egg—to the large number of different cell types seen in adults. In an animal such as the frog, there are likely to be about 150–200 different types of cell in adults. These changes in *structure* are likely to result from *biochemical* changes in the cells. The process of how cells become different—*cellular differentiation*—can thus be thought of as a mechanism that allows certain biochemical *processes* to occur in some cells but not in others (Figure 3).

cellular differentiation*

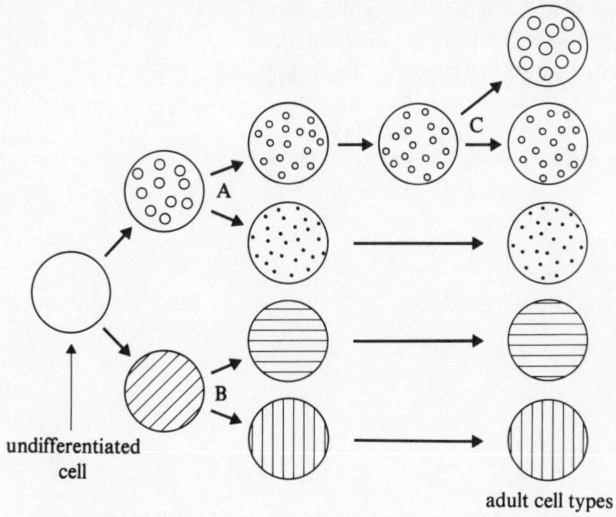

FIGURE 3 The process of cellular differentiation. An undifferentiated cell (*the unfilled circle*) can develop in a number of ways to become an adult cell. What controls development is whether the biochemical processes necessary to produce 'circles' or 'dots', for instance at A, are switched on at the appropriate time.

At points A, B and C in Figure 3 the cell is presented with a 'choice'. For instance, at A the cells produced after division can manufacture either 'circles' or 'dots'. What governs this choice, and how is it made? By 'choice' we are not suggesting that a cell has a mind; we are using this word as a way of describing the situation facing a cell at a branch-point. Why does it go one way or the other? Having gone one way, how does the cell change direction? We ask these questions now, but you will discover by studying these Units that they have not yet been fully answered.

For the moment we can take two features of development as axiomatic.

1 The zygote contains information for making all the different types of cell characteristic of that organism. A zygote is therefore said to be *totipotent* (from the Greek for 'all powerful').

totipotent*

2 The process of cellular differentiation is regulated or controlled.

There must be sufficient information in the zygote[2] to generate all the cell types present in the adult organism and any intermediate cell types that arise during development. This information is by definition 'genetic information'. For example, the information in a fertilized frog's egg for making a frog is passed from one generation of frogs to the next. As you know, this genetic information is called the genotype of the organism. Within a given species of organism the genotype will be broadly the same; a frog is always a frog. However, minor differences in genotype can occur, leading, for example, to differences in eye colour or, somewhat more dramatically, differences between male and female. During development, there is an interaction between the genotype of the organism and the environment in which it is expressed. This interaction yields the final adult organism; that is, it produces the phenotype of the organism. So different environments can affect development in different ways, though only within fairly narrow limits: a fertilized frog's egg cannot be made to develop into a rabbit, or even for that matter into a different species of frog. Drastic alterations in the normal environment tend to be lethal or at least to arrest normal development severely so that an adult, fertile organism is not attained. You will, of course, already be familiar with this concept of genotype–environment interaction from Unit 11; but consider what we mean by environment, when we are looking at the differentiation of a particular cell or a group of cells in an embryo. The environment of this cell (or group of cells) includes not only the external environment of the embryo as a whole but also the immediate environment produced by the surrounding cells within the embryo. Such cell–cell interactions are very important in development; they are considered in detail in Unit 14.

Cellular differentiation is obviously controlled, as is growth. The production of the correct cell types in the correct numbers (let alone the correct positions) is hardly likely to be random. The timing of cellular differentiation is also regulated: production of different cell types happens in a specific order and at specific times during development. This *temporal control* of the use of the genetic information present in the zygote is seen dramatically in organisms that show metamorphosis, that is those with a larval stage. For example, a butterfly egg must contain information for making its larva (the caterpillar) and then a butterfly from this caterpillar. All the changes in cell type that this demands must occur in the correct time-sequence. So, taking our two axioms together: cellular differentiation depends on a controlled use of the genetic information present in the zygote. The use of this genetic information is (in turn) influenced by the environment. 'Signals' from the environment affect the genetic information and may lead to some changes in the cells—and changed cells imply a change in environment for neighbouring cells.

temporal control

Objective for Section 1

Now that you have completed this Section you should be able to:

★ outline two basic assumptions needed to approach the study of cellular differentiation in a logical way.

2 The genetic material in cells

In this Section we consider the basic premise that all the cells of one organism have the same genetic information. This is an expansion of the material you have covered briefly at Foundation level.

2.1 The observational approach

The various specialized cells of a multicellular organism have arisen from one zygote, which must have contained *all* the genetic information to produce these cells; the zygote is totipotent. Thus, differentiated cells could have arisen in the following way:

1 Each different cell type has only the particular genes[3] it needs (Figure 4, scheme A).

2 Each different cell type has *all* the genes for every different cell type but uses only one particular set (Figure 4, scheme B).

3 Some combination of schemes A and B.

Without doing any biological experiments it is possible to get an idea of which scheme is more likely. When somatic cells divide by mitosis the genetic material (associated with the chromosomes of eukaryotic cells), which has previously replicated, is distributed equally to each daughter cell[4]. Thus each new cell carries an identical set of chromosomes—that of the parent cell.

□ Would a simple analysis of the number of chromosomes in each cell enable you to conclude that all cells were totipotent?

■ No. Although the number of *chromosomes* appears to be constant, and they look similar, you cannot say whether the genetic information coded for in the DNA is similar.

Another way of looking at this problem is to measure the amount of DNA in the nucleus of cells from different tissues of the same organism (Table 1).

The variation between the four tissues is within the limits of experimental error. These data support the idea that the amount of genetic material is constant between cells, that is scheme B, and studies of the chromosomes of various insects

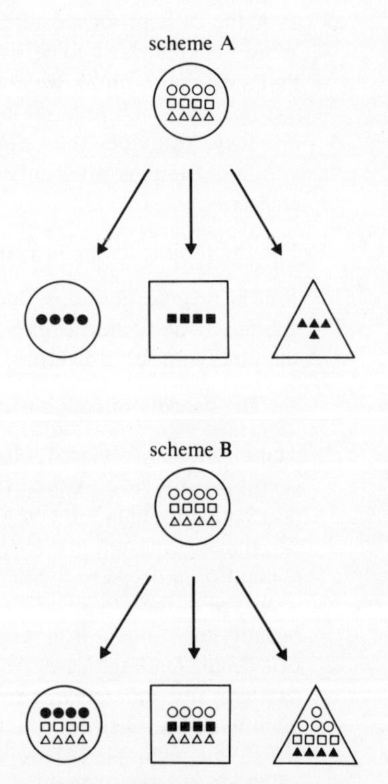

FIGURE 4 The distribution of copies of genetic information. Types of cell are represented by the large circle, square or triangle. Different genetic information is represented by the small circles, squares or triangles. In scheme A only part of the zygote's information is received by each cell. In scheme B each cell receives all the information, but only a part of this is expressed.

TABLE 1 The amount of DNA in the nuclei of cells from various bovine tissues

Tissue	Weight of DNA per nucleus/g × 10^{-12}
thymus	6.6
liver	6.4
pancreas	6.9
kidney	5.9

provide further evidence for this. Some insect tissues have cells with greatly enlarged chromosomes and a DNA content that can be several thousand times that found in normal chromosomes. These chromosomes show a distinct banding pattern when particular stains are used to show up the DNA. Each chromosome has its own banding pattern. In such cells there is no cell division, but there *is* duplication of the DNA, so we find that this chromosome has 1 000 or so identical strands of DNA arranged in parallel (Figure 5).

FIGURE 5 Chromosome number 3 from Malpighian tubules in *Chironomus tentans*, a midge.

You can see that there is an ordered pattern to this particular chromosome; this is seen in all the cells of one tissue. We cannot go into details, but there is strong evidence that the banding is due to the arrangement of DNA within the chromosome.

FIGURE 6 Chromosome number 3 from gut cells in *Chironomus tentans*.

□ Compare Figure 5 with Figure 6, which shows the *same chromosome* but from cells in a *different tissue*. What do you see?

■ There is a remarkable similarity in the banding patterns of these chromosomes. This strongly suggests that the genetic material in the cells of the two tissues is similar.

Because the DNA has been replicated many times, banding is clearly visible in these insect chromosomes, and this makes them especially convenient to study. But, with special staining techniques, banding can be shown in the chromosomes of cells of a wide range of organisms. In all the organisms studied, the patterns of banding appear to be constant for the same chromosome from different tissues.

From these three pieces of information: (a) that the number of chromosomes is constant after the division of somatic cells, (b) that the amount of DNA in different somatic cells is constant and (c) that the appearance of DNA patterns within the chromosomes is constant between cells in different tissues, we are led to the general conclusion that the genetic material is constant between the cells of multicellular eukaryotic organisms.

Although, in practice, this conclusion is generally proved correct, you should be cautious about assuming that it applies to *all* organisms. You could rightly assume that cell division is a basic process in all organisms and thus expect the number of chromosomes to be the same in all somatic cells. However, as far as the amount and position of DNA in chromosomes go, you have been given data from only *two* organisms, the cow and a midge larva. (In Section 2.3, we shall mention some organisms these conclusions do not hold for.) Apart from being confined to a few organisms, this 'observational' approach has further problems. All we have looked at is the gross aspect of the genetic material in different cells; it is not clear whether minor differences in genetic material exist. However, the major weakness of this approach is that we cannot tell if the genetic material can be *expressed* in all the cells. In the next Section we consider the idea that even the most highly differentiated cells of an organism contain the genetic information that can code for all the proteins, and hence determine the structure and function of any other cell of that organism. This is the concept known as totipotency.

2.2 Totipotency

totipotency*

If scheme B in Figure 4 is a correct representation of the distribution of genetic information during cellular differentiation, one could make some interesting predictions about the potential behaviour of differentiated cells. If we could take a single cell from a differentiated tissue and place it in a chemical medium in which the cell could divide, it might be possible for the cell to be reprogrammed and for cells generated by many subsequent divisions to differentiate and, eventually, yield a whole organism similar to that from which the original cell was obtained. This would depend on two conditions being fulfilled:

1 Scheme B is correct.

2 The correct 'environment' for the single differentiated cell to simulate normal development is provided. (We should not expect this to be easy—after all, it is not particularly easy even to get zygotes to develop outside their normal environment; as yet only limited development of mammalian embryos has been achieved 'in a test-tube'.)

Of course, if scheme A is correct then no amount of tampering with the experimental environment could yield a whole organism from a single differentiated cell.

In practice, one can isolate single cells or fragments of tissue from a wide variety of animal organs and stimulate the cells to divide in chemical media. This division is often quite efficient, and large numbers of cells can be grown and maintained in this way. The composition of the required media varies from tissue to tissue and is usually determined in the laboratory by a process of trial and error. Cells have been kept or 'cultured' in this way for many years and the technique is commonly known as *tissue culture*. As the name suggests, cells isolated from a particular tissue usually divide to give cells like those in the tissue from which they were originally derived. For example, individual muscle cells grown in culture can, given appropriate conditions, eventually fuse together to give a muscle-like tissue which can even contract! Although changes do take place in cells in tissue culture, there is seldom any organized differentiation into other cell types. The aim of most investigations using tissue-culture techniques is to maintain cultures with all the characteristics of the tissue from which the cells originated, so the media have been developed mainly to prevent changes in the phenotype of cells. Thus, all that these studies do is to show that the phenotype of differentiated cells is stable—as we might expect because the genetic material is constant. Such studies do not allow us to choose between schemes A and B, because the inability to grow a whole organism from one differentiated cell may merely mean that the second condition—the correct environment—has not been satisfied. However, the story does not end there. Experiments with tissue culture in plants have provided rather more positive results.

tissue culture

2.2.1 Plant cell culture

plant cell culture

In 1948 Caplin and Steward were studying the growth of small pieces of root tissue from carrots in culture media[5]. Growth was slow because cell division was very limited. In trying out various chemical media they found that cell division could be greatly accelerated by adding coconut milk! This is not as crazy or as lucky as it may seem. Coconut milk is a liquid forming the food-storage tissue that is present around and important to the development of embryos and young plants. Additions to the medium, of for example certain plant hormones known to be important in normal plant development, further improved the growth rate of the carrot tissue, but nevertheless it did not differentiate to give a whole plant. Some time later Steward and his associates turned their attention to carrot cultures based on isolated carrot cells rather than pieces of tissue. Once again using the coconut milk medium, good cell division was achieved; a suspension of about 50 cells per cm^3 could give rise to a suspension of about 10 000 cells per cm^3 in 2–3 weeks. Often, however, after a cell division the two daughter cells did not separate, and after the next cell division their daughters did not separate either. In this manner nodules of cells arose. These structures, which Steward called embryoids, in turn gave rise to plantlets and eventually to mature, normal, carrot plants. This outstanding experiment could be repeated on a variety of plants and, furthermore, in some species of plants, cells taken from any one of a variety of organs (stems, root, etc.) could all give rise to whole plants.

The topic of tissue culture in plants will be covered in more detail in Unit 29, but from this brief description you should realize that these experiments imply that there is something special about small groups of cells. Although the total concentration of cells is high there are few *individual* cells—most cells are found as clumps. What is interesting (and as yet unknown) is why some clumps will develop into embryoids whereas others become broken up when the culture is constantly shaken. One factor that could be important is the relative concentration of auxin[6]. More recent work using *single cells* derived from cells cultured in conditions that ensure all the cells have the same genetic material (i.e. cells produced by cloning; Unit 11, Section 2) has shown conclusively that single cells derived from mature plant tissue can develop into whole plants. Thus, differentiated cells in a mature plant, like the original zygote, appear to be totipotent! In plants, at least, there seems to be firm evidence supporting scheme B. Does the same apply to animals? Experiments involving transplants of nuclei between cells have provided an answer to this question.

2.2.2 Nuclear transplants

It is possible to remove the nucleus from the fertilized egg of an amphibian and then watch the subsequent development of the enucleated egg. The usual result is that the egg dies!

□ Why does this happen?

■ For growth and metabolism the cell needs a whole range of proteins, particularly the enzymes involved in metabolic processes; these are coded for by messenger RNA (mRNA)[7]. Thus if the nucleus, and hence the DNA, is removed, metabolism will stop and the cell will die because there is no template to produce mRNA. Removal of the nucleus will also affect cell division in the developing egg.

However, it is possible to replace the nucleus with one from another cell. Two Americans, Briggs and King, exploited this technique of *nuclear transplantation* in the 1950s and were able to transplant nuclei from frog cells at different stages of development into a frog egg.

nuclear transplantation*

□ From these experiments, what results would you expect if scheme A (Figure 4) were operating?

■ Nuclei from cells of advanced embryos would be unable to support proper development of the egg because such cells would contain only a fraction of the total genetic material necessary to support normal development.

The method and results of these experiments are summarized in Figure 7.

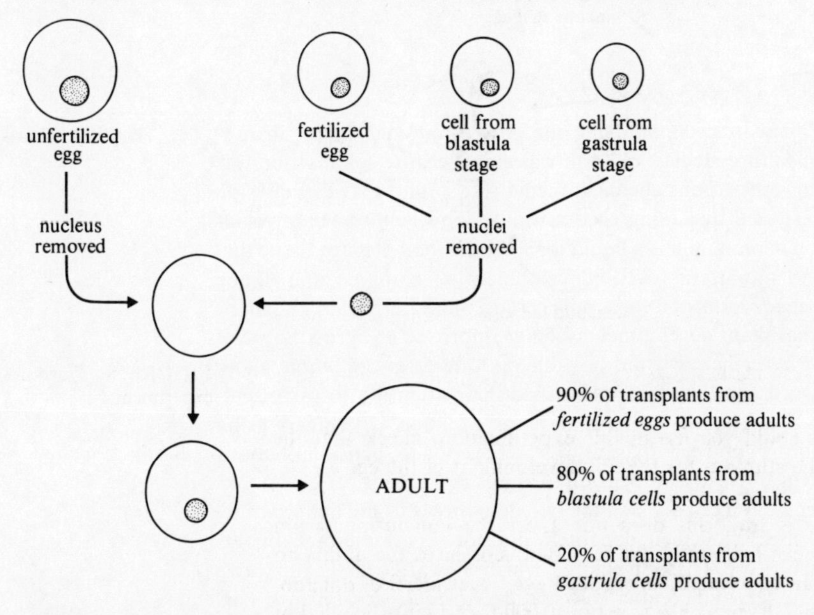

FIGURE 7 Briggs' and King's transplantation experiments.

□ Do the data in Figure 7 support scheme A or B of Figure 4?

■ The transplantation of nuclei from later developmental stages is less likely to produce adult frogs. This suggests that these nuclei may have 'lost' some genetic information. However, as 20 per cent of the transplanted nuclei from the gastrula stage do produce adults, these nuclei at least *must* have all the genetic information to produce an adult frog. The fact that *some* transplants are successful supports scheme B, but the fact that there is a far higher failure rate when older nuclei are transplanted, compared with nuclei from cells at an earlier stage, gives some weight to scheme A.

You will gather that the results of these experiments are not conclusive. There are a number of points of experimental technique that may explain these results.

To transfer the nuclei, a very fine glass tube—a micropipette—is used to suck up a single cell. In the process, the cell membrane is broken and the experimenter ends up with the cell nucleus in the micropipette. This nucleus is then introduced into an egg that has already had its nucleus removed in the same way. Thus, the egg membrane has been broken by the micropipette *twice* to effect the removal and then the introduction of the nuclei. This may well affect the ability of the egg to develop into an adult organism. Another factor is that the donor cells become smaller with age. Nuclei from older cells are more likely to be damaged during transfer because it is more difficult to separate them from the rest of the cell.

ITQ 1 With this further information in mind, which scheme, A or B, do you think Briggs' and King's experiments support?

Answers to ITQs begin on p. 49.

The main question that these experiments did not resolve was whether the nucleus from a *fully differentiated* cell could direct the normal development of the cytoplasm of an unfertilized egg. This was answered in 1962 with transplantation experiments[8] conducted by Gurdon in Oxford. He was able, using a slightly different technique, in which the host nucleus is inactivated by ultraviolet light, and a different species of amphibian—*Xenopus laevis*, to achieve complete development of nuclear-transplant eggs using nuclei from the epithelial cells of tadpole intestine. These cells are considered to be fully differentiated. However, as Figure 8 shows, the failure rate was very high—only 10 per cent of eggs were able to develop, and only 1 per cent of the transplants developed into adult amphibians.

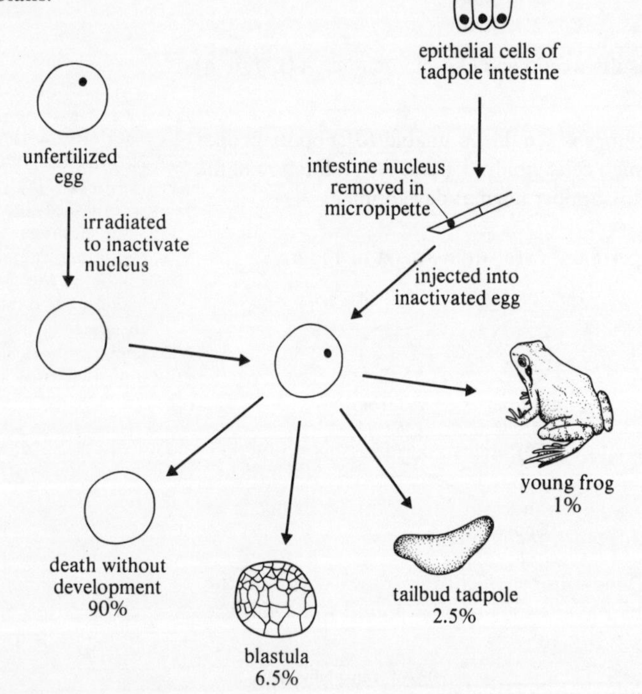

FIGURE 8 Gurdon's transplantation experiments.

ITQ 2 What controls could you use in this experiment to check that the transferred nuclei are the stimulus for the full development of the egg?

Although the success rate is low, this does not affect the conclusions from Gurdon's experiments. Nuclei from fully differentiated cells have the ability to direct the development of an enucleated, unfertilized egg—the nuclei of differentiated cells are thus totipotent. Even if only, say, 1 in 10 000 transplants resulted in adult frogs, you could still make this deduction. It is fairly clear now from current work that the introduction of chromosomal abnormalities is the most important factor in the failure of nuclear transplants. Somatic cells normally divide infrequently (once in 14–28 hours) and it takes about 7 hours for their chromosomes to be replicated before cell division. However, the nucleus starts to divide in the host egg within one hour of its transfer. Thus, in a large number of eggs, nuclear divisions will proceed *before* all the chromosomes have been replicated. Successful transplants will come only from nuclei in which chromosomes have finished replicating. This could explain why there is a higher success rate with nuclei from earlier stages of development—cells at these stages are dividing faster and there is a greater chance of introducing a nucleus in which chromosomal replication *has* occurred.

At this point, it would be tempting to state that nuclear transplant experiments have proved the totipotency of fully developed nuclei, and move on to discuss the control of this genetic information. However, there are a few organisms in which some nuclei are not totipotent.

2.3 Totipotency—exceptions to the rule

You should note that, to the best of our knowledge, successful nuclear transplants of fully differentiated cells have been made in only a few species, most of which are amphibians. This could be because scientists have not tried hard enough with other organisms, or there could be particular factors associated with the amphibian egg. Some workers in Russia (1979) claim success with transplanting the nuclei of fish eggs. At any rate, the evidence in animals is from a very narrow range of organisms. The work with plant tissue culture is more encouraging because Steward's experiments have been repeated with a great diversity of plant species.

We saw in Section 2.1 that the chromosomal material is *constant* in somatic cells, though there are a number of organisms in which a change in the amount of chromosomal material is the rule rather than the exception. A good example is *Parascaris equorum*, a parasitic nematode that has two large chromosomes in the zygote; these break up during cleavage of the egg to form many small chromosomes. Only the cells that will end up producing the germ tissues (testis and ovary) retain the complete set of chromosomes. The remaining (somatic) cell nuclei lose part of their chromosomes to the cytoplasm.

This is illustrated in Figure 9 which is based on drawings made at the end of the last century by one of the founders of investigative embryology, Theodor Boveri.

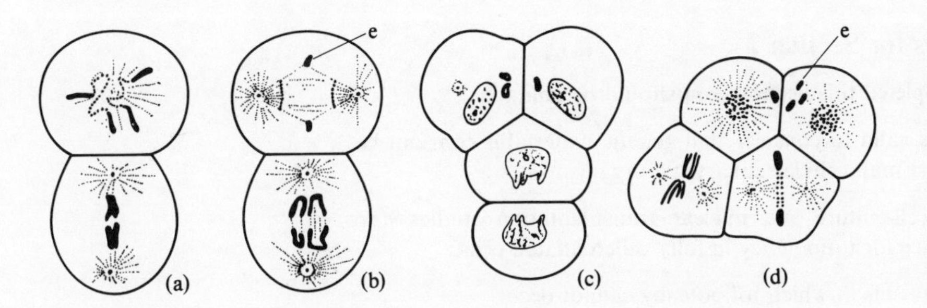

FIGURE 9 Partial loss of chromosomes in the somatic cells of *Parascaris*. (a) The beginning of the second cleavage. (b) A later stage of the same cleavage. The chromosomes are becoming fragmented and their ends (e) lost in the cytoplasm. (c) The four-cell stage, viewed from the animal pole. (d) Later in the four-cell stage. The chromosomes are becoming fragmented and their ends lost.

Breakage of chromosomes occurs early in development and after this no further losses occur, thus the amount of genetic material in somatic cells is relatively constant after the early divisions. Similar changes in genetic material occur in some insects, for example the gall midges (order Diptera).

Another phenomenon, an increase in the number of chromosomes, occurs in the tissues of some organisms. Mammalian liver cells are often found to have more chromosomes than other tissues.

You should appreciate that these examples show that there is no such thing as a general rule of totipotency. However, because the cells of most organisms *are* totipotent, we need to explain how it is possible to select which parts of the genetic material are expressed at different times. We can make two points from the work described in this Section.

1 In plants, in some differentiated tissues, under certain conditions, separated cells can divide and then develop into a whole plant.

2 When the nucleus of a differentiated cell from an amphibian is placed into egg cytoplasm it can express its full genetic information to give an adult organism.

ITQ 3 From points 1 and 2, what are the possible influences on the differentiated cell that maintain it in a differentiated state?

Hence, points 1 and 2 provide support for the hypothesis that, in most species, differentiated cells have nuclei that are totipotent—that is they contain the same information as the zygote. How can we explain why differentiated cells *are* specialized when they all have the same genetic information? In Section 1 we outlined the idea that the changes in differentiated cells are the result of biochemical changes. You know that genes code for mRNA, which then codes for specific proteins[9]. Thus, control over either the *transcription* of DNA to mRNA or the *translation* of mRNA into protein (or some combination of translation *and* transcription) could allow genetic material to be expressed in different ways at different times. In Section 3 we shall consider the role of the cytoplasm in controlling the expression of the genetic material in the nucleus.

control of transcription*
control of translation*
differential expression of genes*

Summary of Section 2

Most organisms have identical genetic material in all their cells. This conclusion follows from observations that the number of chromosomes and the amount of DNA in the somatic cells of most organisms are constant.

These observations alone do not allow us to conclude that the genetic material can be expressed in all cells. Experiments with plant cell culture show that for a large number of plant species, isolated, differentiated cells or clumps of cells are able to develop into mature plants under appropriate conditions. Nuclear transplantation experiments show that, for at least a few animal species, the differentiated adult nucleus, when introduced into egg cytoplasm, is able to direct the 'normal' development of the egg into an adult organism. The plant cell-culture experiments also suggest that cell–cell contact could be an important factor influencing development.

Totipotency cannot be common to all cells because some organisms show loss of chromosomes during development. However, given that totipotency is a general feature of most cells, differentiation must involve control over the expression of the genetic material. This could be by control of the transcription of DNA and/or control of the translation of mRNA.

Objectives and SAQs for Section 2

Now that you have completed this Section, you should be able to:

★ give reasons why it is valid to consider that genetic material is constant between the cells of most multicellular eukaryotic organisms.

★ describe how plant cell-culture and nuclear transplantation studies have provided good evidence for totipotency in fully differentiated cells.

★ give examples of organisms in which totipotency cannot occur.

★ outline the main factors that could maintain cells in a fully differentiated state.

To test your understanding of this Section, try the following SAQs.

SAQ 1 (*Objective 3*) Which of the following statements provides the best support for the concept of constancy of genetic material between all the cells of an organism? Discuss each statement in turn.

(a) Chromosome number is constant in all the cells of an organism, except in the germ cells which have half the number of chromosomes found in the somatic cells.

(b) Banding patterns in some insect chromosomes are virtually identical when examples of the same chromosome taken from different tissues are compared.

(c) The weight of DNA is very similar in all the somatic cells of the organism.

(d) Chromosome number, the appearance of the chromosomes and the weight of DNA are identical in all somatic cells of the organism.

SAQ 2 (*Objectives 4 and 5*) Which of the following statements are *true* and which are *false*?

(a) In the majority of organisms, the somatic cells of the adult contain the same genetic material and so all fully differentiated cells have the potential to be totipotent.

(b) Cell culture work with plant material has shown that it is the general environment in which the cells find themselves rather than direct cell–cell contact that is crucial for differentiation.

(c) The success of nuclear transplants in amphibian cells has provided the best evidence that fully differentiated cells are totipotent.

(d) The failure of some nuclear transplants in amphibians can be explained by the damage to some nuclei, and thus the loss of some genetic information, occasioned by the technique.

(e) Contact between cells and the interaction of the cytoplasm with the nucleus of the cell are the two main factors that appear to control the expression of the genetic material in the nucleus.

(f) Gall midges and frogs are two examples of organisms whose cells are totipotent.

3 Nucleus and cytoplasm

From the nuclear transplantation experiments discussed in Section 2 it is clear that the interaction of nucleus and cytoplasm may be important in the differentiation of cells. In this Section, we shall consider how investigations of cell hybrids and the alga *Acetabularia* have shown the importance of the interaction of nucleus and cytoplasm to development.

3.1 Cell hybrids

It has been known for some time that in cultures of an animal tissue two or more cells occasionally fuse together forming a *multinucleate cell*. In the early 1960s it was shown that the number of such *cell fusions* could be increased dramatically in a cell culture if certain viruses[10] were present. If the virus is irradiated with ultraviolet light before being incubated with the cells, its nucleic acid is damaged and it cannot multiply when added to the cell culture. Nevertheless, such an inactivated virus can still promote cell fusion because it is probably the protein coat of the virus that has this property. The cells produced can be of two types:

multinucleate cell

cell fusion

1 Cells that contain two or more separate nuclei. These cells do not divide successfully.

2 Cells that initially contain two nuclei which fuse and then divide during mitosis to give two daughter cells, each with a large nucleus containing the complement of chromosomes of both parent nuclei. Such 'double-sized nucleus' cells can divide and grow.

In 1965, Henry Harris and his colleagues at Oxford demonstrated that it was possible, using viruses, to fuse together cells from *different* animal species to give multinucleate cells. Some examples are shown in Figure 10.

FIGURE 10 Photomicrographs of the results of some cell-fusion experiments: (a) tetranucleate cell; (b) binucleate cell. H = HeLa. E = Ehrlich.

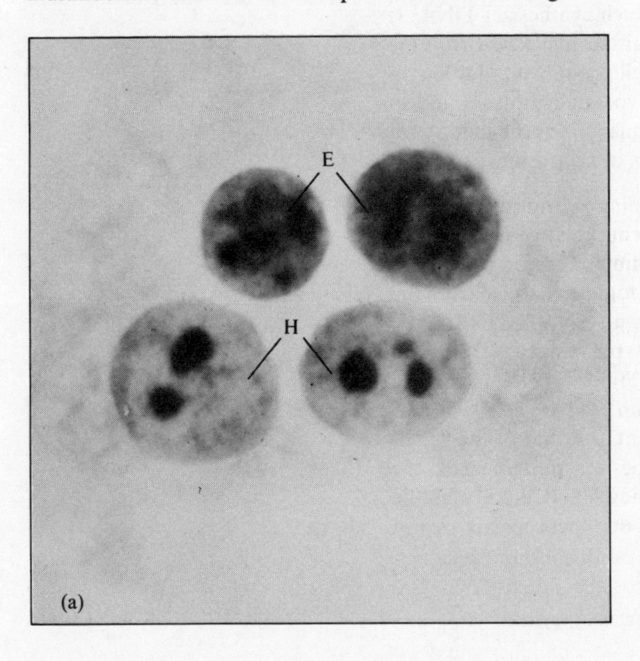

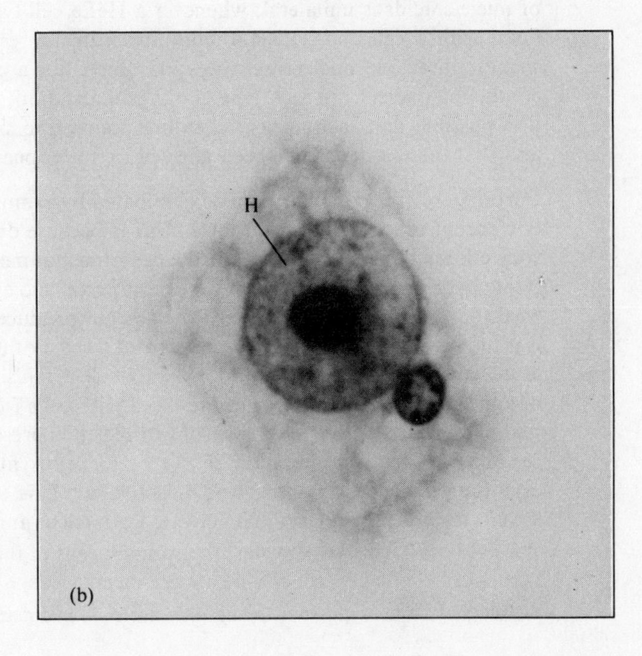

Figure 10a shows a tetranucleate cell derived from two *HeLa cells* and two *Ehrlich* tumour *cells* (HeLa cells are cells cultured from tissues derived from a human tumour. Ehrlich cells are a form of mouse tumour cell.) Because the nuclei are different in appearance they are readily distinguishable. In Figure 10b you can see a binucleate cell derived from one HeLa cell and one hen *erythrocyte* (a red blood cell*).

HeLa cells
Ehrlich cells

erythrocyte

□ What serious problem would arise if this cell fusion work was repeated with plant material?

■ Plant cells have a cellulose wall that would prevent the fusion of the cell membranes.

It is possible, however, to free plant cells from their cell walls by using enzymes to digest them. With such 'wall-free' cells or *protoplasts*, the fusion of cells of different species has been achieved.

protoplasts

As with cells from a single species, binucleate cells derived from two different species can undergo nuclear fusion to give cells with a single, large, nucleus containing the chromosomes characteristic of both species. Such cells can divide and grow. These cell-fusion experiments allow the relationship of nucleus and cytoplasm to be investigated. For example, the nucleus of a rabbit macrophage cell (a specialized, white blood cell) synthesizes RNA but not DNA. When this cell is fused with a HeLa cell whose nucleus carries out both activities the resulting hybrid contains two separate nuclei, each of which now synthesizes RNA *and* DNA. The results from a whole series of similar experiments are summarized in Table 2.

TABLE 2 Results of cell-fusion experiments

Cell type		Synthesis of RNA	Synthesis of DNA
HeLa (human tumour cells)	single cells	+	+
rabbit macrophage		+	–
rat lymphocyte		+	–
HeLa–HeLa	hybrid cells	+ +	+ +
HeLa–macrophage		+ +	+ +
HeLa–lymphocyte		+ +	+ +
macrophage–macrophage		+ +	– –
macrophage–lymphocyte		+ +	– –

NB Each + or – represents the activity of one nucleus.

Examine Table 2 carefully. These results suggest that the regulation of the synthesis of nucleic acids is unilateral: whenever a HeLa cell (which synthesizes DNA) is fused with a cell that does not, *both* nuclei in the resulting hybrid do so. This suggests that one nucleus activates the other. But a cell hybrid, in addition to containing nuclei from each type of cell, contains both types of cytoplasm, and so it is possible that each nucleus is still responding to signals from the cytoplasm and that these signals have been affected by the process of cell fusion.

Fortunately, this possibility can be eliminated by examining hybrid cells in which one parent cell is a hen erythrocyte. This is because during cell fusion the erythrocyte loses all its cytoplasm into the experimental medium. Thus, in a multinucleate hybrid of a hen erythrocyte and another cell, no cytoplasm derived from the erythrocyte is present. The erythrocyte nucleus produces no RNA or DNA, but if it is fused with any active cell, such as a HeLa cell, both types of nuclei (HeLa and erythrocyte) in the hybrid produce these nucleic acids. Similar results have been obtained when hen erythrocytes are fused with cells from a wide variety of animals. In each case, only one type of cytoplasm is present, not that of the erythrocyte. Yet, in each instance, if the other cell synthesizes nucleic acid, the erythrocyte nucleus in the hybrid is activated. This shows that there must be signals from the cytoplasm that activate the nucleus and that these signals are not specific to a particular species.

* Note that avian erythrocytes retain their nuclei, unlike mammalian erythrocytes.

ITQ 4 Are the data on DNA synthesis relevant to studies of the differential transcription of genes?

The important conclusion from this work is that the synthesis of RNA can be initiated or restarted by non-specific signals in the cytoplasm.

3.2 *Acetabularia*

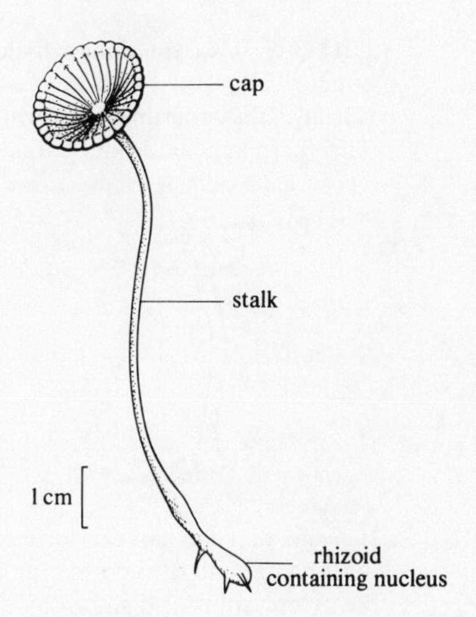

FIGURE 11 *Acetabularia.*

You might have got the impression so far that the only important work on the expression of the genetic material has been done in the last 20 years. We shall now look at a system first studied in the 1930s.

Acetabularia is a marine, green alga. It is particularly useful as an organism for developmental studies because it consists of one very large single cell (Figure 11).

The special shape of the cell and position of the nucleus make it relatively easy to detach and transplant the nucleus. This is done by cutting the stalk just above the rhizoid (which contains the nucleus) and transplanting the stalk to a new rhizoid.

Two species of *Acetabularia* exist with distinctly different cap structures. The cap is a specialized part of the cell and produces the gametes. A classic experiment involved removing the cap from individuals of the two species and then exchanging nuclei by the stem-grafting technique. Figure 12 illustrates this experiment.

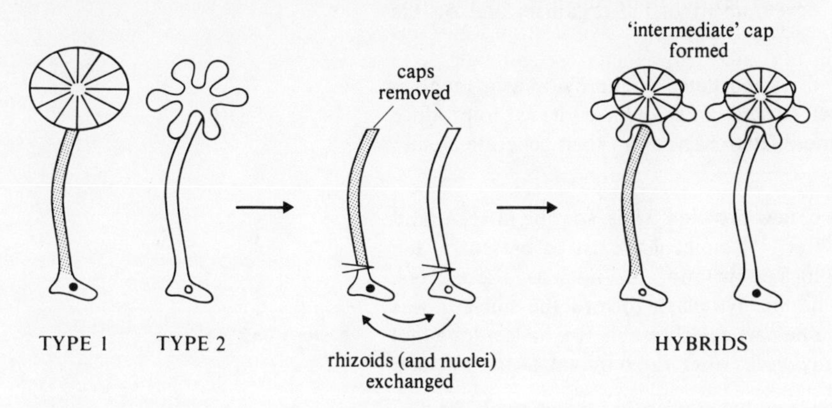

FIGURE 12 Cap-grafting experiment on *Acetabularia.*

When new caps developed they had an appearance intermediate between the two forms.

□ What do you deduce about the roles of cytoplasm and nucleus in *Acetabularia* from this experiment?

■ It seems that signals from both the nucleus and the cytoplasm may be involved because the nucleus from one type of cap in conjunction with cytoplasm from the other type produces a cap with characteristics of both.

If these hybrid caps are removed and the stems are regrafted onto their original rhizoids, the stalks regenerate caps with characteristics of the 'original' nuclei (Figure 13).

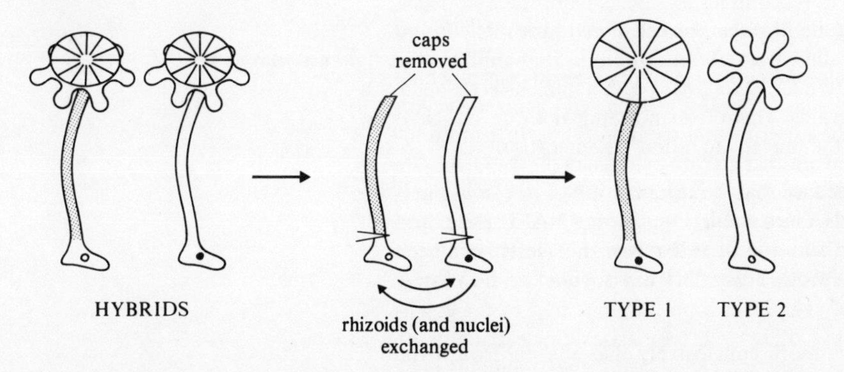

FIGURE 13 Cap-grafting experiment: re-grafting the stems.

It appears from this experiment that although the cytoplasm has some effect—hence the hybrid cap—the nucleus also has a role in determining the structure of the cap. It is only when the right combination of rhizoid, stalk *and* nucleus exists that the normal type of cap can be formed.

ITQ 5 Can you think why hybrid caps were formed in the first experiment?

Consider the experiment illustrated in Figure 14.

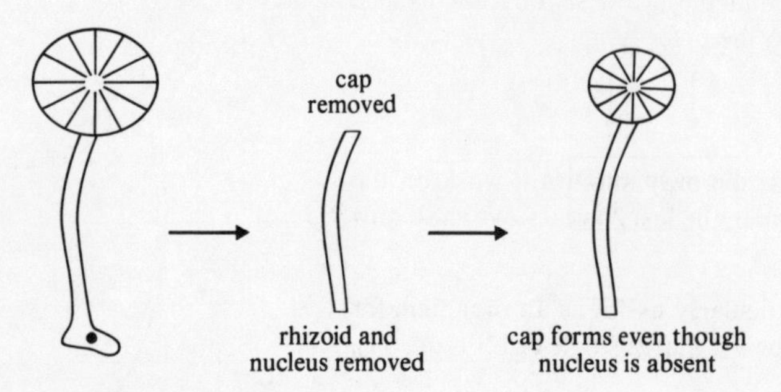

FIGURE 14 The formation of a cap in the absence of a nucleus.

It seems that the cap can form in the absence of a nucleus; thus the cytoplasm must contain all the information necessary for cap formation. The importance of the cytoplasm is also shown by its effects on nuclear division prior to the formation of gametes. Normally, the nucleus in the rhizoid divides to give large numbers of daughter nuclei. These migrate to the cap where the gametes are formed. If, however, the cap and stalk are removed, no nuclear division occurs. The nucleus divides only after a new stalk and cap have formed. It is clear that the cytoplasm and cap have major effects on the development of the organism and on the activity of the nucleus.

A further experiment that indicates the importance of the cytoplasm involves cutting an *Acetabularia* cell in half before stalk and cap have formed to produce one half with a nucleus and one without. *Both* halves will then generate a stalk and cap.

Cap formation involves the synthesis of new proteins. Thus, specific mRNA (and non-specific ribosomal and transfer RNA[11]) molecules must be present in the cytoplasm. Because a cap can be formed without the nucleus, these RNA molecules *must* have been present in the cytoplasm before the nucleus was removed. This *stored mRNA* must also be particularly stable (i.e. have a long life) because cap formation can occur many weeks after the removal of the nucleus.

stored mRNA

Thus, for this organism we could suggest an hypothesis that the control of differentiation is not by the specific *transcription* of DNA in the nucleus but by the control over the *translation* of 'long-life' mRNA. What evidence is there that mRNA *is* long-lasting? Consider two experiments (Figure 15).

EXPERIMENT 1

Following on from the last experiment: if the two halves of a young *Acetabularia* cell are treated with an enzyme, *ribonuclease*, that breaks down RNA (the enzyme is added to the medium around the cells) neither half generates a stalk and cap because both the existing (long-life) and newly synthesized mRNA are broken down. When the ribonuclease is removed from the medium, regeneration occurs only in the half with a nucleus.

ribonuclease*

EXPERIMENT 2

Another experiment involves cutting the *Acetabularia* cell close to the nucleus and then treating the two parts with *actinomycin D*, an antibiotic that inhibits the transcription of DNA. The regeneration of the fragment with a nucleus is stopped, although the other fragment regenerates. There does not appear to be sufficient mRNA in the fragment containing the nucleus to 'allow' regeneration.

actinomycin D*

From these experiments you could deduce that the regeneration of *Acetabularia* is dependent on the transcription of DNA into stable, long-life mRNA. These experiments, though, are fairly crude; the addition of an enzyme that destroys all the RNA in a cell may affect the cell in other ways. It is also unclear how these enzymes can actually enter the cells!

A more elegant investigation shows more conclusively that the control of the production of different proteins is organized in the cytoplasm rather than in the nucleus; that is, control is exerted at the level of translation rather than transcription. *Acetabularia* produces three different phosphatase enzymes—these can be

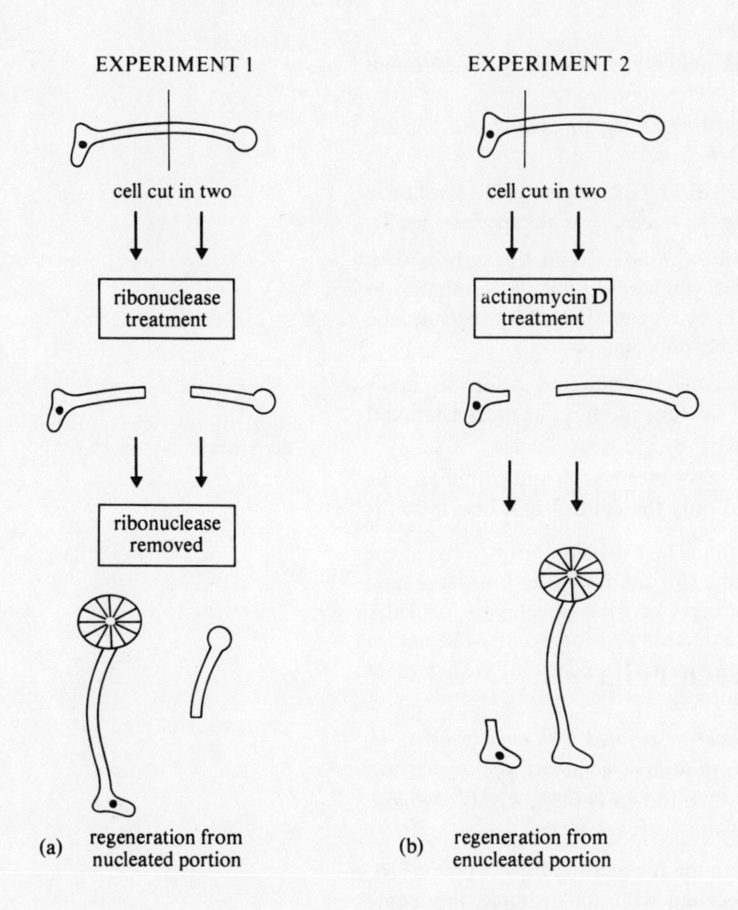

FIGURE 15 Ribonuclease and actinomycin D experiments.

distinguished by their different optimum pH values. In the normal, nucleated cell these are produced in a precise sequence before cap formation. This precise sequence is also seen in cells that have had their nuclei removed. Thus, there must be specific translation in an ordered manner of the long-life mRNA in the cytoplasm.

This work on cell hybrids and *Acetabularia* is very significant. In Section 2, we concentrated on the idea that differentiation involves the controlled expression of genetic information coded in the DNA of the cell. In this Section we have arrived at the conclusion that the cytoplasm itself can control the expression of the DNA, either at the level of transcription (i.e. by acting on the DNA to control the production of RNA; cell hybrid experiments) or, as in *Acetabularia*, at the level of translation. These points will be taken up again in Sections 8 and 9.

Summary of Section 3

The work of Henry Harris on animal cell hybrids has provided a valuable experimental system for studying the interactions of cytoplasm and nucleus. The experiments show that the synthesis of nucleic acids can be stimulated in nuclei that normally do not synthesize DNA or RNA if the nucleus is associated with the cytoplasm of a cell that does synthesize nucleic acids. Because the cytoplasm of unrelated species can stimulate the synthesis of nucleic acids it appears that non-specific cytoplasmic 'signals' can control the expression of DNA. *Acetabularia*, as well as showing nucleo–cytoplasmic interactions, provides evidence that there is control over the *translation* of mRNA as well as over the *transcription* of DNA.

Objectives and SAQ for Section 3

Now that you have completed this Section, you should be able to:

★ describe what (and how) cell-fusion experiments contribute to our understanding of nucleo–cytoplasmic interactions.

★ outline experiments that provide evidence for the control of the translation of cytoplasmic mRNA in *Acetabularia*.

To test your understanding of this Section, try the following SAQ.

SAQ 3 (*Objectives 7 and 8*) Which of the following statements are *true* and which are *false*?

(a) Cell-fusion experiments have shown that there must be *cytoplasmic* signals, which could control the transcription of DNA.

(b) In a rabbit macrophage–hen erythrocyte hybrid, both nuclei would synthesize DNA because the signals which may activate the nucleus are not species-specific.

(c) If the cap of one species of *Acetabularia* is removed and the stalk is then transplanted to a rhizoid of a different species, you would expect a cap similar to that normally produced by the new rhizoid to be regenerated. This is because the nucleus is the main factor determining cap morphology.

(d) Treatment of *Acetabularia* with the enzyme ribonuclease before a cap is formed stops cap formation. This shows that cap formation is due to translational rather than transcriptional control.

(e) The sequential production of specific proteins in cells with and without nuclei provides good evidence against the idea that only the control of transcription is involved in cellular differentiation.

4 Biochemical differences between cells

This Section deals with the biochemical changes associated with development—in particular, changes in enzyme proteins. You will need some knowledge of enzymes (Units 6 and 7) and there are some references to metabolic pathways, which you met in Unit 8.

Differentiation of cells is to do with changes in the proteins that make up the cell. These include both structural proteins associated with, for example, the membranes around and within the cell and the enzyme proteins involved in metabolism[12].

The synthesis of macromolecules and the interconversion of small molecules is dependent on enzymes. Thus, any change in the structure and activity of the cell during development which leads to differentiation must at some point be determined by the type and amount of particular enzymes within the cell. Consider Figure 16.

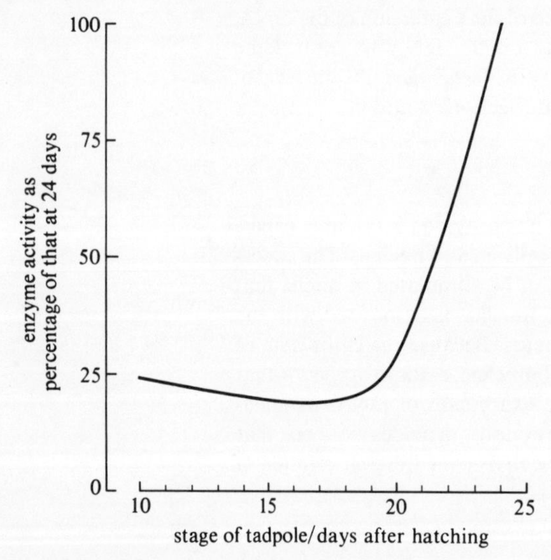

FIGURE 16 The changing activity of the enzyme arginase in tadpole liver.

We can measure a change in the amount of an enzyme, arginase, produced by the liver when tadpoles turn into frogs and leave their aquatic environment (21 days after hatching). Most organisms need to remove waste nitrogen derived from the breakdown of proteins. The simplest way is to excrete ammonia. However this is highly toxic, so unless the organism is in water, where the ammonia can simply diffuse away, a non-toxic product must be synthesized. In terrestrial organisms such as the frog this is urea. Arginase is one of the enzymes in the urea cycle, a metabolic pathway that converts amino acids to urea. In the frog there is a sudden rise in the level of arginase when the frog adopts a terrestrial habitat. (You will read more about nitrogen excretion in Units 22 and 23.)

Another example of biochemical changes during development is shown in studies on liver enzymes in rats.

Figure 17 shows changes in the activity of three enzymes in rat liver from 5 days before birth to around 24 days after birth. Enzyme A is involved in the polymerization of glucose molecules to form glycogen. This carbohydrate is crucial for the fetal rat because it provides an energy store for use after birth. The activity of

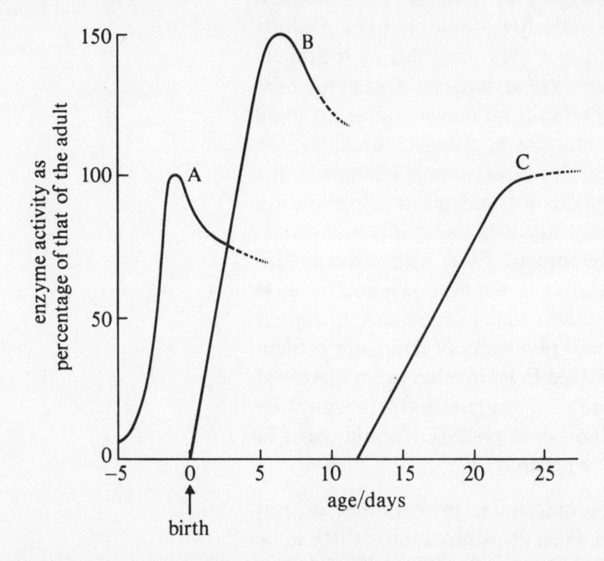

FIGURE 17 The activity of some enzymes in rat liver.

enzyme A declines about 1 day before birth, after which the activity of enzyme B rises rapidly. This enzyme is important in helping to maintain a constant level of blood sugar by converting an intermediate in the citric acid cycle (oxaloacetate) to phosphoenolpyruvate. Recall from Unit 8 (Section 2.1) that phosphoenolpyruvate is an intermediate in the glycolytic pathway. If the level of this metabolite is increased, there will be a tendency for the glycolytic pathway to work backwards and thus synthesize glucose. Enzyme C breaks down excess tryptophan. There is very little of this amino acid in rats' milk. When a rat's diet changes, the level of tryptophan rises and enzyme C is needed to break it down. You should not worry about the details. The reason for mentioning this work is to emphasize that during development definite changes in proteins (in this case enzymes) seem to occur.

Now, you know that:

1 Different cells contain different proteins.

2 Different cells of one organism contain the same genetic information.

3 DNA contains the information for synthesizing proteins—one gene codes for a particular polypeptide chain via mRNA[13].

Thus, there are a number of hypotheses, each of which could account for cellular differentiation.

HYPOTHESIS I

Different cells transcribe and translate different genes, thus producing different proteins and, hence, various kinds of cells.

HYPOTHESIS II

Different cells transcribe the same genes, but translate different mRNA molecules.

HYPOTHESIS III

All genes are transcribed and translated in every cell but the rate of degradation of particular proteins varies in different cells.

What we wish to examine now is whether any of these hypotheses (or any combination of these hypotheses) explains the changes seen during cellular differentiation. In other words, we wish to answer two questions:

1 What is the relative importance of the degradation and synthesis of specific proteins?

2 Is control over cellular differentiation exercised through transcription of the DNA or translation of the mRNA?

When trying to answer such complex questions it is generally desirable to use an experimental system in which the phenomena being examined can be controlled by the experimenter. Cellular differentiation is complex and occurs as part of an even more complex process, development. For this reason, most studies of what factors control changes in the cellular levels of proteins have been carried out not on developing systems but on adult organisms in which changes can be brought about in a few proteins under carefully controlled conditions. The rationale behind the experiments is that the mechanisms underlying these changes in adults will probably be similar to those underlying changes that occur during the differentiation of cells. For example, changes in the levels of several enzymes in the liver of a rat can be effected by changing the animal's diet from one containing a small amount of glucose to one containing a large amount of glucose. Similarly, the injection of certain hormones into an animal can lead to dramatic changes in the levels of specific proteins in specific tissues. The dietary changes or injections are of course controlled by the experimenter and the effects occur rapidly and do not depend upon long periods of growth and development. Even with these advantages, it is often very difficult to establish the answer to the first question because synthesis and degradation both occur in the tissues of higher organisms. In experiments, it is difficult to distinguish between these two ways of changing protein levels: for example, if an increase is seen in enzyme E, is this due to an increased rate of synthesis of E or a decrease in the rate of degradation? It would be convenient to use a system in which only synthesis *or* degradation could occur at any one time so that each could be examined separately.

Unlike higher organisms, bacteria, which are unicellular prokaryotes, do not exhibit a significant rate of protein degradation when growing rapidly; this can be demonstrated readily by growing bacteria in the presence of radioactively labelled amino acids. During growth the bacteria synthesize proteins and these will carry the radioactive label. The labelled amino acids are then removed from the medium and replaced by unlabelled ('cold') amino acids. The subsequent loss of radioactivity from the labelled, cellular proteins (because they are broken down and resynthesized using non-radioactive amino acids) is found to be very slow. Under some circumstances bacteria do, however, exhibit changes in the level of certain proteins and these changes must be due to increased protein synthesis. For this and other reasons (as you will see) most of what we know about the control of protein synthesis has come from studies on bacteria.

Summary of Section 4

From the work on changes in the levels of liver enzymes in the frog and rat, we can deduce that cellular differentiation may involve the ordered synthesis and degradation of proteins.

Some hypotheses are proposed to account for differentiation. These involve the differential transcription and translation of genes, the differential breakdown of protein, or some combination of both of these systems. There are problems in investigating these hypotheses in eukaryotic organisms. There is little breakdown of protein in bacteria, so they are good organisms to use to provide a model for the control of protein synthesis during development.

Objectives and SAQ for Section 4

Now that you have completed this Section, you should be able to:

★ give one example of biochemical changes associated with a developmental process.

★ outline some hypotheses that could account for differentiation.

★ outline the problems of designing experiments to test the relative roles of protein synthesis and protein degradation in developing organisms.

To test your understanding of this Section, try the following SAQ.

SAQ 4 (*Objective 10*) After injecting substance X into a rat there is a progressive increase in the level of X-ase, an enzyme that destroys X in the liver. Other proteins do not alter in level. The time course of this increase is shown in Figure 18.

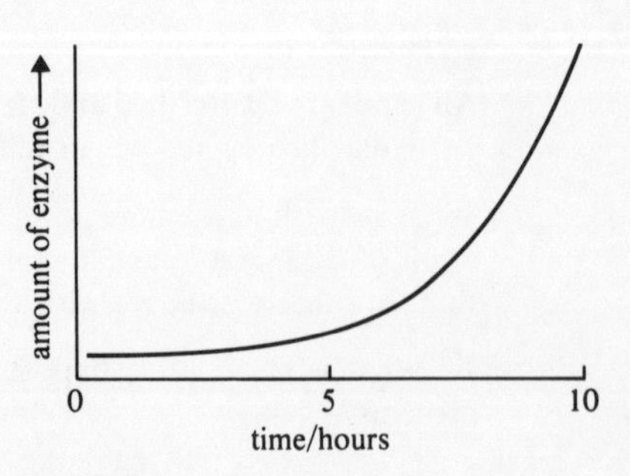

FIGURE 18

Which of the following statements are *true* and which are *false*?

(a) The increase in X-ase could, in principle, be due to an increased rate of protein synthesis. This could be demonstrated by injecting radioactively labelled amino acids with substance X and isolating radioactively labelled X-ase several hours later.

(b) If the increase in X-ase is solely due to an inhibition of the rate at which it is degraded, then the increase would not be altered by the addition of an inhibitor of protein synthesis.

(c) The increase in X-ase could be due to an increase in the rate of translation of mRNA for X-ase.

(d) The increase in X-ase could be due to a raised level of mRNA for X-ase in the cell following increased transcription of the gene for X-ase.

5 The control of protein synthesis in bacteria

If you have not already attempted the pre-Unit test on protein synthesis (p. 4), try it now.

5.1 Enzyme induction in bacteria

More is probably known about *Escherichia coli* than about any other single organism. *E. coli* grows when provided with a simple medium comprising NH_4^+ (as a source of nitrogen), SO_4^{2-} (as a source of sulphur) PO_4^{3-} (as a source of phosphorus), small amounts of some other metal ions and a source of carbon. The carbon source can be any one of about 20 different organic compounds. But *E. coli* grows most rapidly on glucose. On this sugar *E. coli* divides about once every 30 minutes when grown at 35 °C. The glucose is converted by the glycolytic pathway, TCA cycle and all the interconnected metabolic pathways to provide the carbon necessary for synthesizing the organic compounds that comprise *E. coli* (Unit 8). All these metabolic conversions require specific enzymes which *E. coli* also synthesizes, and these again require carbon.

Consider, however, what happens when we provide *E. coli* not with glucose but with another source of carbon such as lactose. Lactose is a disaccharide consisting of one molecule of glucose linked to one molecule of galactose via a β-galactoside link. To metabolize lactose, *E. coli* must first break this β-galactoside link by means of a specific enzyme, β-galactosidase. The glucose and galactose formed can then be further metabolized via the glycolytic pathway.

lactose —β-galactosidase→ galactose + glucose

When *E. coli* is grown in a medium containing glucose as the sole source of carbon, the cells contain little β-galactosidase. If some of these *E. coli* cells are removed from the glucose medium and placed into a medium in which the sole source of carbon is lactose, there are some dramatic changes in the level of β-galactosidase in those cells (Figure 19).

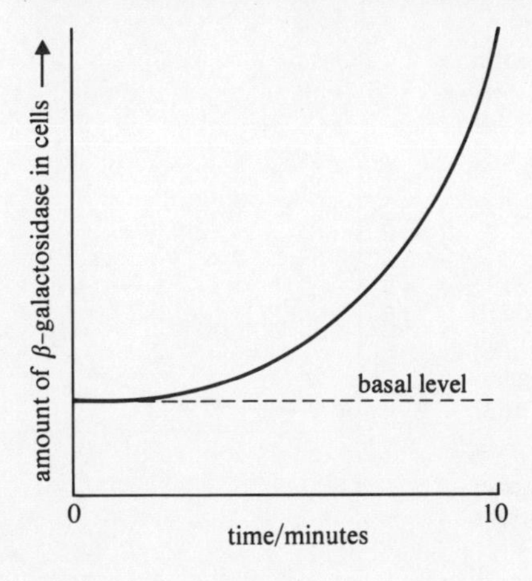

FIGURE 19 Changes in the amount of β-galactosidase in *E. coli* cells. The glucose-grown cells were transferred to lactose at time 0. The original level of β-galactosidase (at time 0) is called the 'basal level'.

The general name for this phenomenon is *enzyme induction**. In this example, the enzyme β-galactosidase is induced by its substrate, lactose, which is termed the *inducer*. The *E. coli* cells that are able to manifest this increase are said to be inducible and once the increase has occurred the cells are said to have been *induced*.

enzyme induction*

inducer*

The advantages of enzyme induction are clear. If a cell makes a high level of an enzyme only when the substrate on which that enzyme acts is present, this represents a considerable economy to the organism. Thus, if lactose is removed from induced cells (and replaced by, say, glucose) or broken down by those cells, no further increase in the level of β-galactosidase occurs. The β-galactosidase present does not break down because there is little protein degradation in bacteria. However, if no further β-galactosidase is being synthesized, the level of β-galactosidase per cell decreases at each cell division: as each cell divides to give two, the level per cell drops by 50 per cent, at the next division by a further 50 per cent, and so on. If lactose is added to such cells, induction can occur again, and so on. There is a 'ceiling' to the induction; but with certain analogues of lactose, induction can result in up to a thousand-fold increase over the basal level of β-galactosidase; the induced enzyme can represent up to about 6 per cent of the total cell protein.

This huge increase in the level of β-galactosidase must arise from an increase in the rate of synthesis of this protein. Many different substances are required in the synthesis of any protein. Alteration of the level of any of these could result in a change in the rate of protein synthesis. For example, if an organism suffered a reduction in the level of adenosine triphosphate (ATP), its rate of protein synthesis would be expected to fall because protein synthesis is dependent on energy. The rate of protein synthesis would also fall if any of the 20 amino acids was reduced in level. However, enzyme induction is characterized by the fact that it is specific, which means that in response to a particular inducer only one protein (or sometimes a few related proteins) increases in level. This must mean that some component of protein synthesis that is specific to one protein (or a few particular proteins) must be affected.

□ Name the components that are specific to the production of proteins.

■ 1 The specific genes (i.e. the section of DNA) that provide the information for the sequence of amino acids in the polypeptides.

2 The mRNA molecules copied from those specific genes.

So, enzyme induction could involve two separate steps in protein synthesis. For β-galactosidase, for example, it could involve an increased rate of transcription for the gene for β-galactosidase to give more mRNA for β-galactosidase (called β-galactosidase mRNA), and/or an increased rate of translation of β-galactosidase mRNA to give more of the polypeptides that comprise β-galactosidase. There is also the possibility that an increased level of enzyme activity could be due to the modification of existing polypeptides.

Thus, we have three hypotheses that could account for the induction of β-galactosidase in *E. coli*.

HYPOTHESIS I

There is increased transcription of DNA.

HYPOTHESIS II

There is increased translation of mRNA.

HYPOTHESIS III

There are changes in the activity of existing polypeptides.

Over the last 30 years the lactose induction system has been studied by many scientists. These studies have yielded important results and have allowed a decision to be made about hypotheses I–III. We shall deal briefly with the main lines of the research and how it allowed the lactose system to be analysed.

* Do not confuse this with the term 'embryonic induction', referred to in Units 11 and 14. Enzyme induction always refers to increases in enzymes, embryonic induction to changes in cell types. Unfortunately, there has been a tendency over the years to abbreviate both terms to 'induction'.

5.2 The analysis of the lactose system

Consider the following experimental observations.

1 On the addition of lactose (or some non-metabolized chemical analogues of lactose) to *E. coli* cells, an increase is observed in the level of β-galactosidase and two other enzymes, galactoside permease and transacetylase, necessary for the uptake of lactose into the cell. These increases start within a few minutes of the addition of lactose.

2 Puromycin, a substance that inhibits the joining together of amino acids to form polypeptides, when added to *E. coli* along with lactose prevents any induction of β-galactosidase.

3 If radioactively labelled amino acids are added to *E. coli* along with lactose, and the β-galactosidase subsequently induced is isolated and purified, it is found to be labelled.

□ How do these experimental observations support hypotheses I–III?

■ Observations 1–3 have no bearing on hypotheses I and II because they do not relate to the transcription of the DNA or the translation of mRNA. Thus we cannot as yet dismiss these hypotheses. It is clear though that hypothesis III is ruled out because observations 2 and 3 show that induction cannot depend merely on existing polypeptides. If existing polypeptides were involved you would not expect the inhibition of polypeptide formation to affect induction. In addition, there would be no incorporation of radioactively labelled amino acids into polypeptides because induction would not depend on the synthesis of new polypeptides. Thus hypothesis III is not supported by the evidence.

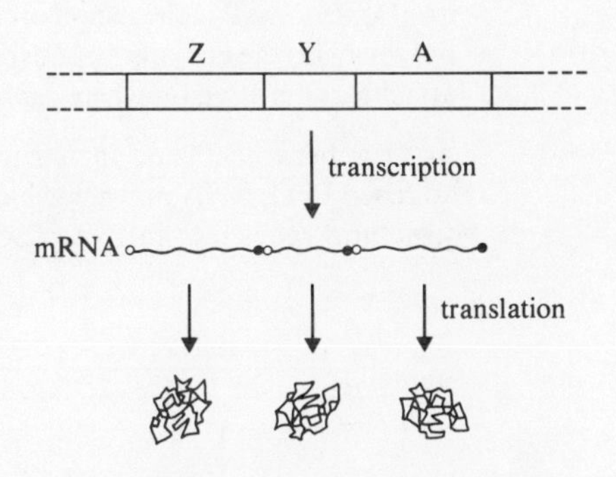

FIGURE 20

Further experimental observations were made to try to distinguish between hypotheses I and II.

4 Using genetic techniques, a map of the relative positions on the DNA of the three genes containing the information for the three enzymes of the lactose system can be constructed. These genes, Z, Y and A, are found close together on the DNA. They are transcribed to give one long mRNA molecule that is translated in, as it were, three regions to give the three enzymes (Figure 20).

5 The half-life of β-galactosidase mRNA is very short, of the order of 1–2 minutes.

6 Some mutants[14] of *E. coli* make a great amount of the three enzymes even in the absence of inducer; these mutants were called *constitutive* because they make the enzymes continually (i.e. as part of their constitution). Using genetic techniques, it can be shown that these constitutive mutations all occur in a separate gene, located near the Z, Y and A genes. This gene is described as a *regulator gene* because it is involved in the regulation of how much enzyme is synthesized. By contrast, Z, Y and A are called *structural genes* because they determine the structure of the three enzymes.

constitutive mutants*

regulator gene*

structural gene*

□ What do observations 4–6 tell us about hypotheses I and II?

■ Observation 4 does not allow us to distinguish between the two hypotheses. Enzyme induction could still be brought about either by increasing the rate of transcription of the three genes (hypothesis I) or by increasing the rate of translation of the mRNA molecules they code for (hypothesis II). It is interesting, though, that the three genes are in some way linked together.

Observation 5 does suggest that control over the transcription of DNA *may* be more important than control over translation in enzyme induction. It is difficult to think of how there could be control over the translation of the mRNA molecules when they are present in the *E. coli* cell for a very short time. However, the half-life of the mRNA might be altered and this would affect how much enzyme was produced.

Observation 6, like observation 4, does not allow us to distinguish between hypotheses I and II. The regulator gene could control enzyme induction by affecting the transcription of the DNA *or* the translation of the mRNA.

These observations are based on experiments carried out by two French biologists, Jacob and Monod. Their studies led to a whole new hypothesis about how protein synthesis might be controlled.

5.2.1 The Jacob–Monod hypothesis

Jacob's and Monod's hypothesis was put forward in the following way.

Jacob–Monod hypothesis*

(i) The regulator gene contains information for making a protein called *repressor*. There are always a few repressor molecules inside a cell. The regulator gene was referred to as the *i gene* because it responded to inducer.

repressor molecule*

i* gene

(ii) Repressor molecules, in the absence of inducer molecules (lactose), specifically inhibit the synthesis of the three enzymes, β-galactosidase, permease and transacetylase.

(iii) Inducer, when present in the cell, binds to repressor.

(iv) Repressor, when bound by inducer, is incapable of inhibiting the synthesis of the three enzymes.

(v) Repressor acts by inhibiting transcription. The inhibition of transcription would mean a rapid 'switching off' of the synthesis of the enzymes because the mRNA has a short half-life (see observation 5).

Jacob and Monod and their co-workers deduced from further genetic studies the existence of two additional regions on the DNA—the *promoter* (P) and the *operator* (O) regions. P is the region on the DNA adjacent to a set of genes where RNA polymerase (the enzyme responsible for copying the DNA sequence into mRNA) attaches. O is a region where repressor can attach if there is no inducer in the cell.

promoter region*

operator region*

As O is between P and the structural genes (Z, Y and A), when a repressor is attached to O, RNA polymerase cannot get to, and hence cannot transcribe, the structural genes into mRNA. These points are illustrated in Figure 21.

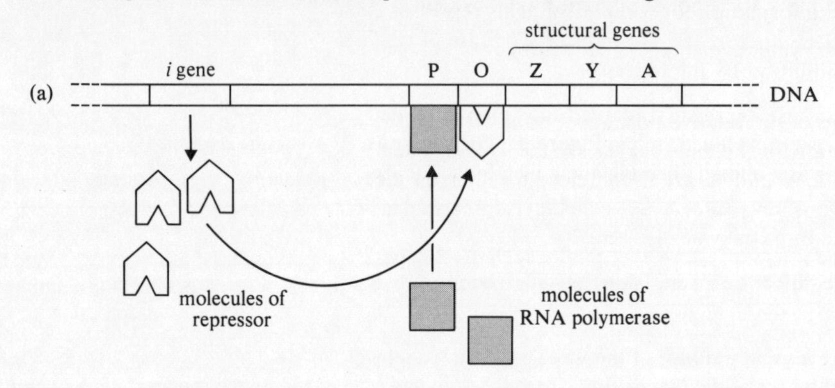

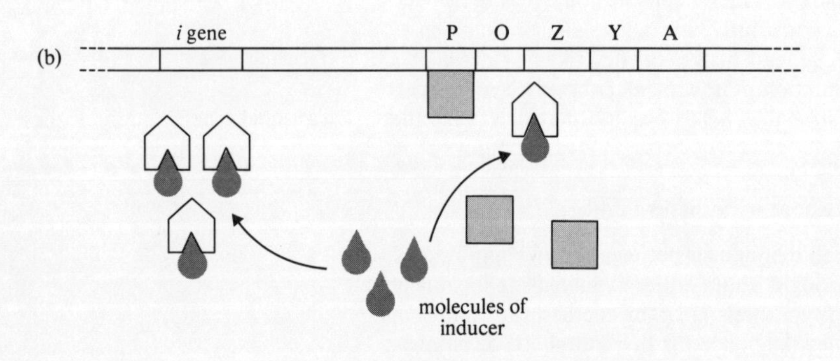

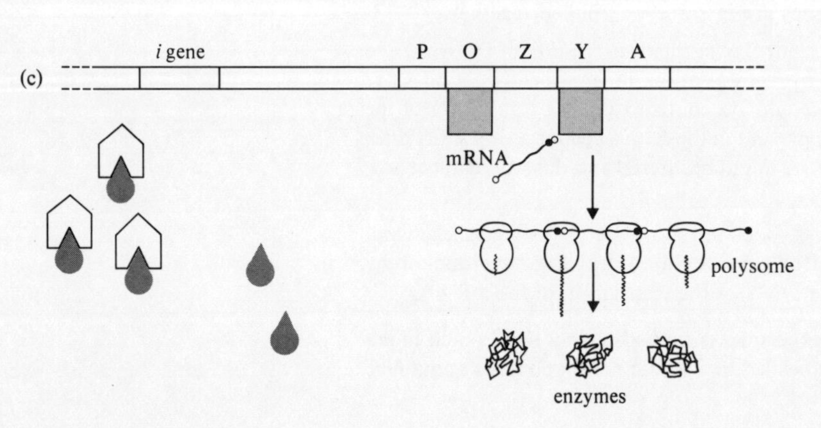

FIGURE 21 The Jacob–Monod hypothesis for enzyme induction.

In summary: repressor (coded for by the *i* gene) binds to O in the absence of inducer and so, because RNA polymerase attached to P cannot reach Z, Y and A, no mRNA corresponding to these genes is made (Figure 21a). When induction occurs, the inducer binds to repressor and then prevents it from attaching to O (Figure 21b). RNA polymerase can now reach Z, Y and A and can transcribe these genes to give the relevant mRNA molecules, which are then translated to give the three enzymes (Figure 21c).

More recent data have shown that the hypothesis advanced by Jacob and Monod in 1961 was correct. The data are as follows:

1 After the addition of inducer to *E. coli* an increased amount of β-galactosidase mRNA can be detected. Inducer seems to cause increased transcription of DNA.

2 Repressor molecules have been isolated. That is, a protein has been isolated that is coded for by the *i* gene.

3 This isolated repressor binds *in vitro* to DNA isolated from the operator region, but not to DNA from other regions. This binding is prevented by inducer.

4 From *E. coli*, it is possible to derive a cell-free system comprising DNA, RNA polymerase, ribosomes, tRNA, activating enzymes and all the substrates of small molecular mass required for protein synthesis. Such a system will synthesize β-galactosidase. This synthesis is prevented by adding isolated repressor.

operon*

Jacob and Monod coined the word *operon* for a system in which the regulation of the synthesis of several enzymes is coordinated and the structural genes for these enzymes are adjacent on the DNA. Thus, for each operon there is a specific regulator gene (e.g. the *i* gene for the lactose operon or *lac* operon, as it is generally called) that contains the information for a repressor. This repressor is specific for the operator region of the operon and the inducer. The rate of transcription is set by the degree of inhibition by the repressor.

enzyme repression*

This Jacob–Monod hypothesis can also be readily adapted to explain a related phenomenon in bacteria, *enzyme repression*, illustrated in Figure 22. The synthesis of certain groups of enzymes that are related through their metabolic roles is

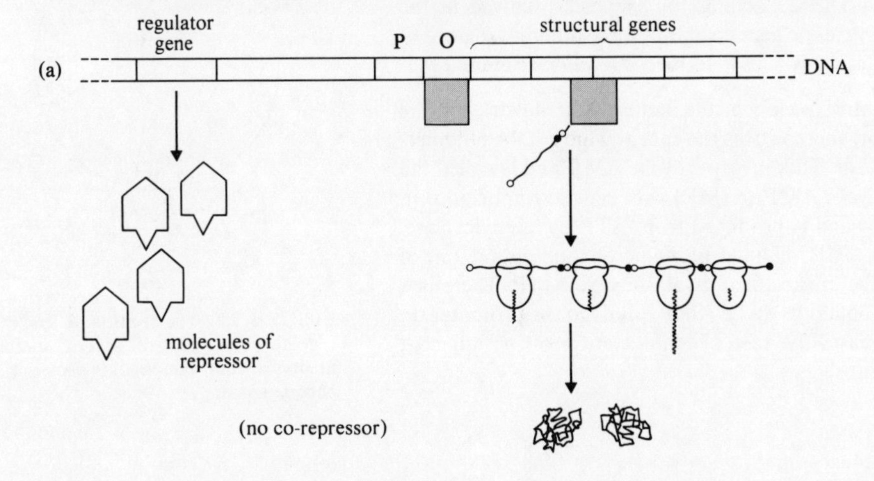

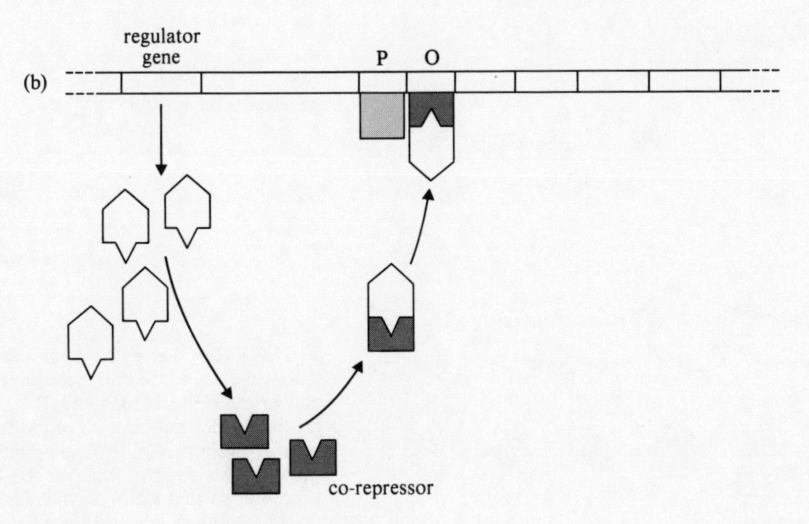

FIGURE 22 The Jacob–Monod hypothesis for enzyme repression.

reduced when certain small molecules, called *co-repressors*, are added to the cells. (This is the reverse of induction in which the small molecule (the inducer) allows enzyme synthesis.) Control of transcription is thought to be exercised by specific repressors. For any one repressible operon there exists a regulator gene that contains information for a specific repressor. This can bind to an operator region, and hence inhibit further transcription, only in the presence of the small molecule co-repressor. (Conversely, in induction, the small molecule, the inducer, prevents the repressor from binding to O.) Existing mRNA, made before the addition of co-repressor, soon decays.

co-repressor molecules*

Not all enzymes are subject to induction or repression; some are made continually. Generally, these are enzymes in central metabolic pathways, such as the glycolytic pathway. There is a constant requirement for these enzymes which is probably geared to the overall growth rate of the cell.

Although it has been proven only for the lactose system in *E. coli*, the Jacob–Monod hypothesis has provided a model for superficially similar instances of induction and repression in many organisms. However, even where induction or repression are shown, the assumption that all these phenomena happen in the same way as the induction of β-galactosidase is somewhat precipitate. The main achievement of Jacob and Monod is not merely in explaining the control of synthesis of β-galactosidase in *E. coli*. More importantly, they have provided a framework within which many studies relevant to the control of protein synthesis in general, and indeed to cellular differentiation, can be initiated. It is this stimulus to further research that was recognized in their share in the Nobel Prize for Physiology or Medicine in 1965.

ITQ 6 Summarize the main points about induction in bacteria and, in particular, about the central features of the Jacob–Monod model.

5.3 Other mechanisms for induction and repression in bacteria

There are several variations on the mechanism proposed in the Jacob–Monod model. For example, one regulatory gene and its repressor may control several operons: this is the situation with genes coding for enzymes involved in the synthesis of the amino acid arginine; these genes occur in three different regions of the chromosome. Such a coordinating system is termed a *regulon* (Figure 23).

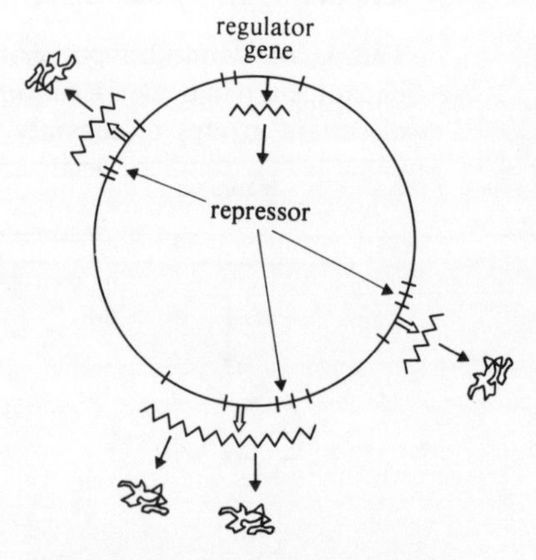

FIGURE 23 The regulon. A single regulator gene controls several operons at sites around the circular bacterial chromosome.

There is also an extra level of control, which for the lactose system works on the promoter rather than the operator, and controls the rate at which RNA polymerase molecules attach to the operator. This involves cyclic AMP and is called the *cyclic AMP effect* (Figure 24). Cyclic AMP (cAMP) is of central importance in many biological control reactions and is produced from ATP by adenylcyclase. (You will learn more about cyclic AMP in Unit 16.) If glucose is abundant, one of its breakdown products lowers the intracellular level of cyclic AMP. There is a *gene-activator protein* that, when bound to cyclic AMP, binds to the promoter on the DNA, and by doing so increases the rate of transcription of the adjacent

cyclic AMP effect

gene-activator protein

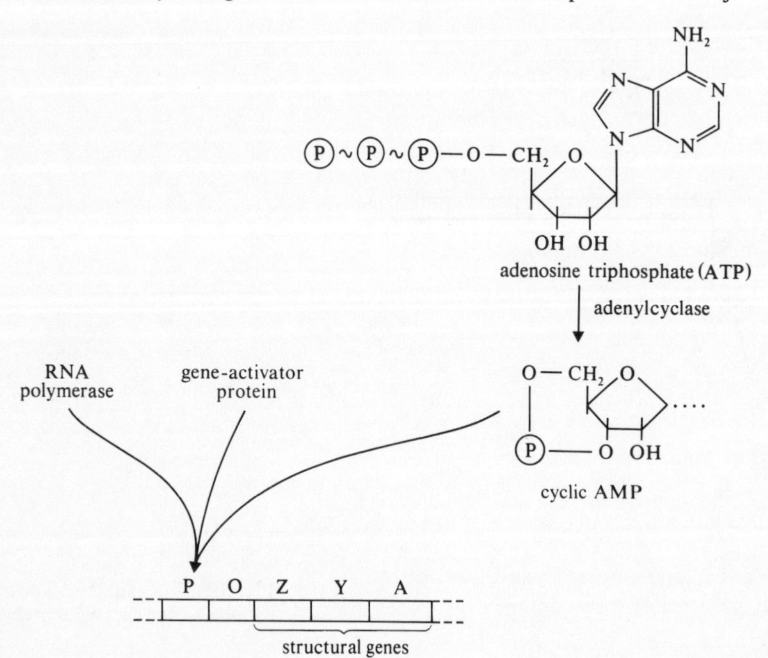

FIGURE 24 The cyclic AMP effect. Adenylcyclase in the cell membrane converts ATP to cAMP. cAMP facilitates the transcription of the *lac* operon by combining with gene-activator protein. This complex changes the promoter so that RNA polymerase can transcribe the gene. Ⓟ = phosphate.

operons. Simply, if the level of cyclic AMP is reduced, then the transcription of the *lac* operon is also reduced. The cyclic AMP effect is not limited to lactose: it also works on operons connected with the use of other sugars. It is therefore an overriding control on the metabolism of alternative energy sources. Other sources are used only if glucose is not available. You should appreciate that induction–repression and the cyclic AMP effect are not the only mechanisms by which transcription of DNA may be regulated; but other mechanisms are beyond the scope of this Course.

5.4 Is all control at the transcriptional level in bacteria?

As we said earlier, the amount of a specific protein could be regulated in two main ways: by regulating the rate of transcription of DNA (i.e. the synthesis of mRNA) and by regulating the rate of translation of mRNA.

So what evidence is there for control over the rate of translation of mRNA? In the lactose system, there are fluctuations in the amounts of enzymes produced under different conditions—which does suggest further control mechanisms. There are fluctuations, for example, in how much β-galactosidase and acetylase is made: these fluctuations can be produced by altering the temperature at which the bacteria are grown. These temperature effects are also seen in the six structural genes of the histidine operon. Gene 1 is translated about three times as frequently as genes 2–6. If a sequence of genes in an operon is transcribed as a single unit of RNA, then the relative amounts of each enzyme produced should be set. That this is not so illustrates that there must be further regulatory mechanisms that act on individual parts of the whole operon, so that one part of the mRNA is translated more often than others; that is there must be control over the *translation* of the mRNA. You should note, though, that the evidence for control of translation is rather circumstantial.

Summary of Section 5

Synthesis of certain bacterial enzymes occurs when substances that are substrates of those enzymes are present in the bacterial cell. This *induction* of the enzyme is rapid and involves the synthesis of new polypeptide chains. Jacob and Monod produced a model for the control of enzyme induction that has been supported by much experimental evidence. This model has the following features:

1 Control is via the regulation of the transcription of DNA.

2 The structural genes for related enzymes that are induced together are adjacent on the DNA.

3 This *operon* is regulated by a specific *regulator* gene that produces a specific repressor molecule.

4 The repressor molecule binds to an *operator* region (adjacent to the structural genes) and prevents mRNA polymerase transcribing the DNA.

5 When the enzyme substrate or *inducer* is present, this binds to the repressor molecules allowing mRNA polymerase access to the DNA. Thus transcription of the structural genes can occur.

There are some variations of the Jacob–Monod model. One involves a regulatory gene controlling many operons and in another example, cyclic AMP may exercise control over the binding of mRNA polymerase to the promoter region.

There is some circumstantial evidence that in bacteria control may also involve translation of the mRNA.

Objectives and SAQ for Section 5

Now that you have completed this Section, you should be able to:

★ outline the Jacob–Monod model of the control of enzyme induction in bacteria.

★ give an example of enzyme induction in bacteria which may involve control at the level of translation rather than at the level of transcription.

To test your understanding of this Section, try the following SAQ.

SAQ 5 (*Objective 12*) Which of the following statements are *true* and which are *false*?

(a) Enzyme repression involves the binding of a repressor molecule to a regulator gene.

(b) Lactose is an inducer of β-galactosidase because it provides the *E. coli* cells with glucose.

(c) An operon usually comprises several adjacent structural genes and an operator region.

(d) The levels of all enzymes in *E. coli* cells are likely to be controlled along the lines of the Jacob–Monod model.

(e) A promoter is a region on the ribosome where tRNA binds.

(f) A regulator gene determines the structure of the enzymes coded for by the operon.

6 The control of gene expression in higher organisms

It would be convenient if we could describe genetic control processes in higher organisms in the same detail as for *E. coli.* However, eukaryotes are more complex for the following reasons: their chromosomes are more highly organized and their DNA is associated with a large amount of protein in what is known as *chromatin* (see the TV programme, *Differential Gene Expression*, and Unit 5, Section 5.4). In addition, mRNA may have a long life in eukaryotes, and may be found prepackaged in an inert form in the cytoplasm, especially of eggs. The presence of a nucleus and nuclear membrane in eukaryotes creates problems of transport between the nucleus and cytoplasm.

chromatin*

You should also note that there are two distinct kinds of regulation in eukaryotes. The first is *short-term regulation*—reversible regulation of the type seen in bacteria. It often represents the cell's reaction to fluctuations in the environment, and involves changes in the amounts or activities of enzymes and hormones. Of equal importance is the *long-term regulation* involved in *developmental* changes.

short-term regulation

long-term regulation

Before we go further, let us consider what is meant by eukaryotic DNA. Look at Table 3.

TABLE 3 The amount of DNA in terms of the number of base pairs per cell in a range of organisms

Organism	DNA/base pairs per cell $\times 10^{-6}$
E. coli	6
Drosophila sp.	120
Paracentrotus lividus (sea-urchin)	800
Homo sapiens	2 800
Triturus (newt)	20 000

□ What do these figures suggest?

■ According to the limited data, eukaryotes appear to have far more DNA per cell than prokaryotes. (However, you should be wary of extrapolating from just these five examples!) The larger amount of DNA per cell in eukatyotes indicates that the DNA is likely to be more *complex.* There is also a great variation in the amount of DNA in the two vertebrates listed. Thus, there is not a direct relationship between the size and complexity of an organism and the amount of DNA its cells contain.

□ Does all this extra DNA in *Triturus* sp. compared with *Paracentrotus* sp. mean that *Triturus* produces 250 times as much messenger RNA?

■ No, because only a proportion of the DNA is likely to be expressed at any one time and also the DNA codes for other RNA molecules—ribosomal and transfer RNA—as well as messenger RNA.

Exactly why particular organisms have more DNA than others is problematical, but it is clear that the large amounts of DNA in most eukaryotic cells mean that the control of the expression of specific coded sequences may be very complicated.

6.1 The complexity of the DNA

Let us now consider the sequences of bases in the DNA that code for RNA. A technique that enables us to do this is *DNA hybridization*. When DNA extracted from cells is gently heated, the weak hydrogen bonds that hold the two strands of the helix together break and the strands separate. Thus, a collection of single-stranded DNA molecules is obtained. When this single-stranded DNA is slowly cooled the strands reassociate, provided the ionic strength and pH of the medium are suitable. The rate at which this reassociation occurs gives a guide to how similar (or dissimilar) the sequences of bases in the DNA are.

DNA–DNA hybridization*

In the actual experiments, the DNA double-stranded 'duplex' is broken into relatively short lengths. Thus, the experimenter is looking at the rate of reassociation of short lengths of DNA to form short, 'hybrid' DNA molecules. However, each length of DNA is a great deal longer than the sequence necessary to code for an RNA molecule.

For example, consider a hypothetical DNA molecule containing only two kinds of base, one strand containing solely adenine bases and the other strand solely thymine bases. You would expect that after being broken up, heated, and separated into two strands, it would reassociate to form hybrid molecules at a very fast rate because there are only two different sequences of bases to recombine—strands made up from adenine and strands made up from thymine. The chance of two complementary sequences meeting each other is thus very high.

□ What would be the rate of reassociation if the DNA had a totally random sequence of bases?

■ Very slow, because there would be a very great number of pieces of DNA, all with different sequences. The chance of the precise two complementary sequences pairing up immediately is thus very low.

Figure 25a shows an idealized DNA molecule consisting of one strand of one base, adenine, and one of the complementary base, thymine. After the DNA has been broken into short lengths and the strands have dissociated following heating, there are only two possible sequences of bases, and so the single strands will

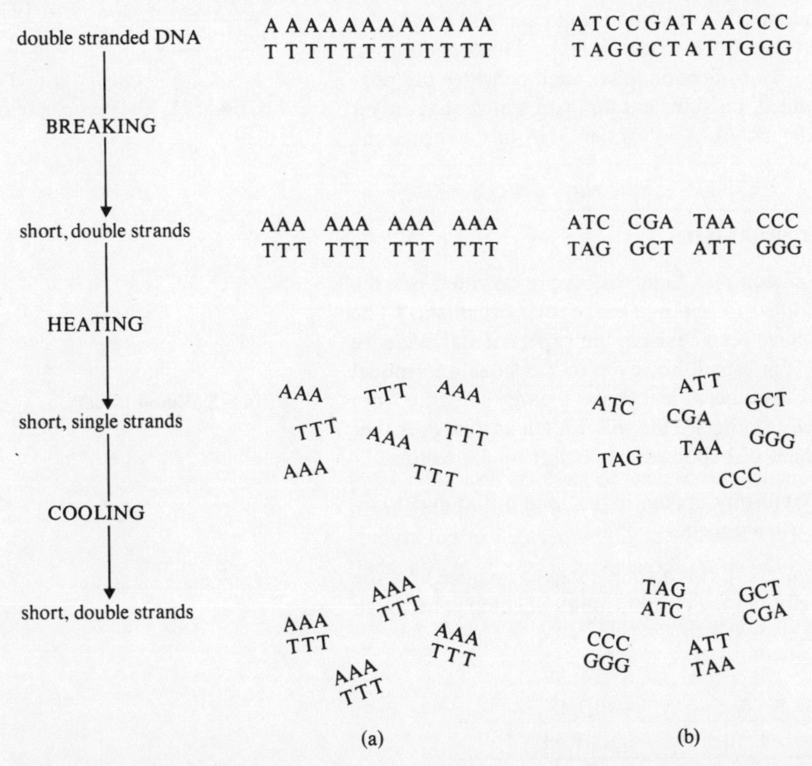

FIGURE 25 The principle of DNA–DNA hybridization.

reassociate quickly when the reaction mixture is cooled. In Figure 25b, the DNA consists of four different bases. You can see that for each of the small single-stranded lengths there is only one possible complementary sequence. The single-stranded lengths are moving about in solution at random, with the result that it will take longer for the complementary sequences to hybridize. You will appreciate that in practice the short, single strands are made up of hundreds of bases—this diagram is a gross simplification!

The rate of reassociation can be plotted graphically (Figure 26). The horizontal axis is obtained by multiplying the original concentration of DNA by the time the DNA takes to reassociate. The vertical axis is the percentage of DNA that has reassociated at any one time. Figure 26 shows such a graph for DNA from mammalian cells and you can see three distinct parts to the curve, each representing a portion of the total DNA concentration.

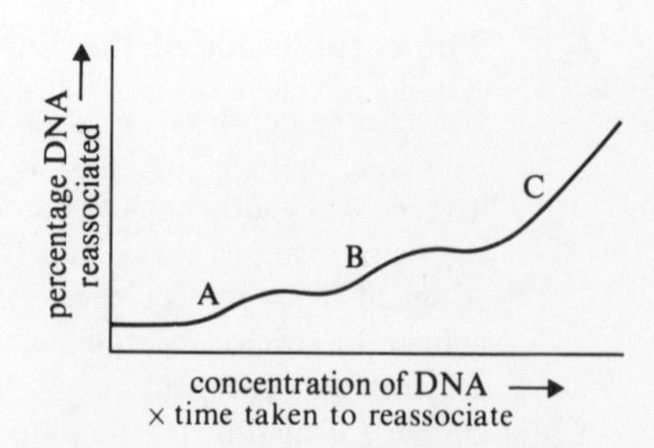

FIGURE 26 The rate of reassociation of DNA. The graph reveals the three fractions (A–C) discussed in the text.

The portion of DNA that reassociates most quickly (fraction A) must have a highly repetitive sequence of bases (a simple sequence). Analysis of this fraction of the DNA shows that it consists of very short, repeated sequences of nucleotides. The significance of these highly repetitive sequences is unclear. The next DNA fraction (fraction B) is known as intermediate repetitive DNA. This fraction has been shown (by studies *in vitro*) to code for tRNA and rRNA molecules. Because this fraction hybridizes faster than fraction C, there must be a number of repeated DNA sequences for these types of RNA. The DNA that codes for histone proteins occurs in this fraction. The last fraction (fraction C), which takes longest to reassociate, consists of what is termed 'unique' or '*single copy*' *DNA* (*scDNA*). This is really a misnomer because the hybridization technique (at least at present) cannot distinguish between one or maybe 15–20 copies. This fraction of DNA contains the sequences that code for mRNA and hence for protein.

single copy DNA (scDNA)*
(Also referred to as SCDNA and ScDNA.)

You may wonder what accounts for the spontaneous reassociation seen in the initial part of the graph in Figure 26. The two DNA strands must be physically close to enable their complementary base sequences to pair up immediately. The explanation of this is probably that this DNA has a palindromic arrangement; that is, the DNA has a complementary sequence of bases reading in both directions from a central point (see Figure 27). When the duplex separates, these sequences on the *same* strand will tend to reassociate to form a structure like a hair-pin. As far as we know now (1980) these sequences have no function.

ATCGATCGAT

TAGCTAGCTA

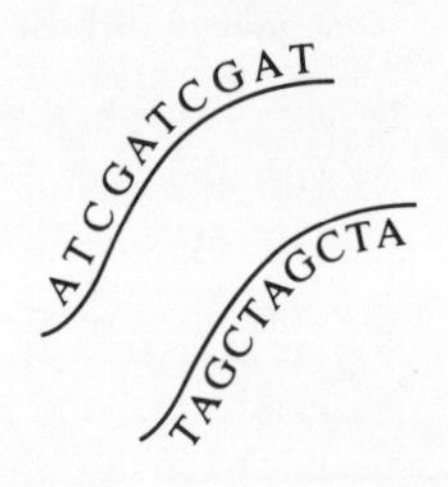

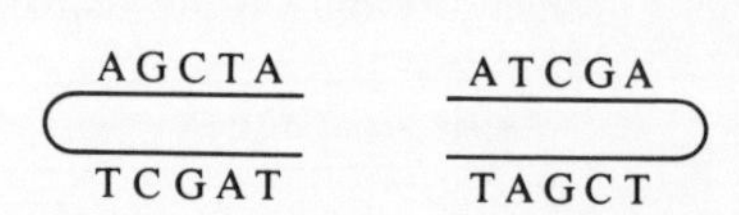

FIGURE 27 Palindromic sequences in DNA.

It is important to realize that all the DNA that codes for RNA in fact accounts for only a small amount of the total DNA in the cell. In *Drosophila* sp. for instance, there is about ten times more DNA present in each cell than the estimated needs, deduced from a rough analysis of the number of proteins and hence genes required. In fact, if one looks at higher organisms, there appears to be even more DNA with no apparent function. In humans the current estimates are that there is between fifty and a hundred times more DNA than one could account for on the basis of the amount of highly repetitive DNA (fraction A) or DNA coding for tRNA, rRNA and histones (fraction B). In Section 9, we shall consider the possible significance of this; for the moment it is sufficient that you realize that only a small fraction of the DNA codes for mRNA. A lot of the DNA has no apparent function!

6.2 A model for control in higher organisms

Given that eukaryotic DNA is more complex than that in prokaryotes, can the basic Jacob–Monod model be modified to apply to eukaryotic organisms? The simple answer is yes—and models have been devised; but experimental evidence has often been difficult to obtain. It is useful, however, to consider one model system, proposed by *Britten and Davidson*, because it shows the way evidence from one biological system can be applied to others. This will not tell us how the other system works but will allow a number of hypotheses to be set up for testing.

Britten–Davidson model*

The basic idea of Britten's and Davidson's model is that *one* initial event can result in *multiple* changes in the activity of genes.

Recall the concept of a regulon (Figure 23). It is a simple step to imagine how the same type of mechanism might function in the chromosomes of eukaryotes. Such a system is illustrated in Figure 28.

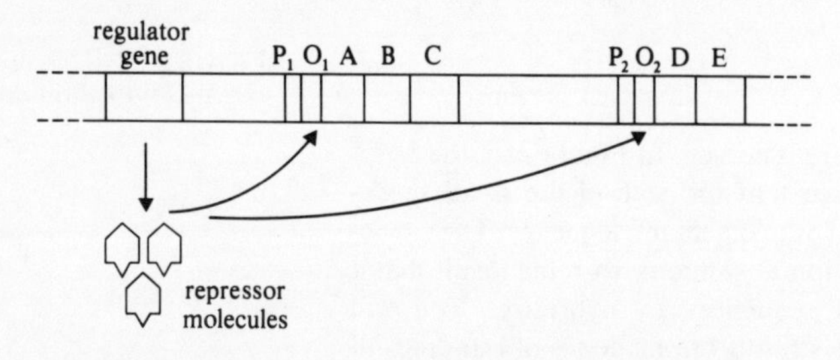

FIGURE 28 A regulator gene controlling two operons.

Britten and Davidson expanded this concept and proposed that there could be genes that reacted directly with molecules such as hormones to 'signal' the start of the required pattern of activity in the genetic material. These so-called 'sensor' genes would control in some way a set of 'integrator' genes, which would produce RNA. This RNA would then associate with a receptor region for each of the operons and allow transcription of mRNA from the 'producer' gene (Figure 29).

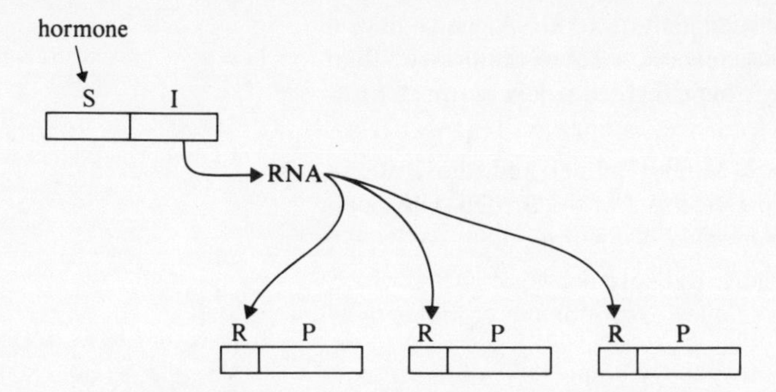

FIGURE 29 The activation of genes according to the Britten–Davidson model for control. S = sensor gene. I = integrator. R = receptor gene. P = producer gene.

There are two major reasons why a more complicated system has to be proposed for eukaryotes. First, the Jacob–Monod model explains how genes can be 'switched off' in the presence of repressor molecules. Thus, you would predict that in eukaryotic cells, a wide range of specific repressor molecules must exist to 'repress' all the genes that are not transcribed in those particular cells. Given the variety of differentiated cells in eukaryotes, this implies a very complex system of repressor molecules, which all have to be synthesized. The Britten and Davidson model gets around this by proposing the existence of molecules that will activate the DNA in response to *specific* stimuli. Producing 'activator' molecules only in response to a stimulus makes a reduced level of synthesis possible and this would be more efficient for the cell. A second reason is the evidence for control over the production of DNA and/or RNA in cell hybrids. In Section 3.1 it was clearly seen that nuclear activity is *activated* by specific cytoplasmic factors.

At present, there is little firm evidence for such control systems in eukaryotic organisms. For instance, RNA 'activator' molecules have not been isolated and, because eukaryotic DNA *is* complex, control regions for each gene have not been identified. However, it is known that mutations that occur at one point in the DNA sometimes affect a number of different genes. This may indicate that the mutation has occurred in the 'sensor' gene or in the genes that produce the 'activator' RNA.

In Section 9 we shall consider more aspects of the control of gene expression with reference to cellular differentiation.

Summary of Section 6

Eukaryotic organisms have much more DNA per cell than prokaryotes. The reason for this is unclear but it suggests that control over the expression of eukaryotic DNA is quite complicated. The technique of DNA–DNA hybridization shows that there are three distinct fractions to the DNA. However, only a small proportion of the DNA codes for mRNA. A possible model for the control of gene expression in eukaryotes is the Britten and Davidson model. This proposes specific 'sensor' genes that respond to signals, such as hormones, which may initiate gene expression. 'Sensor' genes produce molecules that activate a number of different 'producer' genes. Experimental evidence for such a control system in eukaryotes is very limited.

Objectives and SAQ for Section 6

Now that you have completed this Section, you should be able to:

★ outline the ways in which eukaryotic DNA is more complex than prokaryotic DNA.

★ state the principle of DNA–DNA hybridization and interpret observations.

★ give reasons why a more complex method of control than the Jacob–Monod model may be needed for eukaryotic organisms.

★ outline the Britten–Davidson model for the control of gene expression in eukaryotic organisms.

To test your understanding of this Section, try the following SAQ.

SAQ 6 (*Objectives 14–17*) Which of the following statements are *true* and which are *false*?

(a) Eukaryotic DNA, unlike prokaryotic DNA, has many repeated sequences of bases that code for similar gene products.

(b) In hybridization studies, short lengths of single-stranded DNA, which have a high proportion of complementary sequences of bases, will reassociate faster than similar lengths of single-stranded DNA in which there is a low proportion of complementary sequences of bases.

(c) The main difference between the Jacob–Monod model and the Britten–Davidson model for the control of gene expression is that the Britten–Davidson model allows for one event to 'switch-on' a number of operons.

(d) Prokaryotic DNA codes only for mRNA, unlike eukaryotic DNA which codes for a range of RNA molecules. This is one reason for the complexity of eukaryotic DNA.

7 DNA–mRNA hybridization

This Section outlines the techniques used to quantify the different mRNA molecules present in cells at different times. You should appreciate that it is only by being able to measure the production of different kinds of mRNA molecules that we can have a basis for considering the concept of the differential expression of genes, that is the production of different kinds of mRNA molecules during development. The Section is conceptually difficult, so you will need to spend some time on it as you will need to understand it to appreciate the studies described in Section 8. The TV programme, Differential Gene Expression, has an animation sequence on the technique.

7.1 The original experiments

Any type of mRNA will have a particular sequence of nucleotide bases that is complementary to one strand of the particular region of DNA from which it was transcribed, just as each of the double strands of DNA is complementary (Section 6). This property of mRNA can be used as the basis of a system for assaying how much mRNA is present in the cell at any particular time.

If the DNA duplex is heated to separate the two strands and mRNA is then mixed with the single-stranded DNA, on cooling you would expect some of the mRNA to join with sequences of DNA; any one section of DNA will bind only to the mRNA molecule that complements it. As long as there is a large amount of mRNA there will be little reassociation of the single-stranded DNA.

So, if a large amount of a preparation of mRNA (which will contain a number of different mRNA molecules) is added to a small amount of single-stranded DNA, virtually all the regions of the DNA that are complementary to the types of mRNA present will be bound by mRNA. Because there is an excess of mRNA, a number of mRNA molecules will not be able to bind to the DNA—there will be competition for binding between identical mRNA molecules.

When a preparation of mRNA molecules from two different tissues is used in this experiment, it is possible to find out if these mRNA molecules are all of one type by measuring the amount of competition between them for the DNA. It is necessary to be able to recognize the mRNA from each tissue. This can be done by radioactively labelling the mRNA molecules in one of the tissues.

□ How could this labelling be achieved?

■ If radioactively labelled uracil—a base specific to RNA—is injected into an animal, after some time RNA that incorporates the labelled uracil will be synthesized. Thus, when the RNA is extracted from the tissues of this animal it will be radioactively labelled.

Consider what would happen if you mixed labelled mRNA from one tissue with DNA from the same tissue. The mRNA would, under appropriate conditions, bind to the DNA. The resulting double strands would be radioactively labelled. Suppose you then added increasing concentrations of unlabelled mRNA (produced from the same tissue in a different animal of the same species). You would find that this would compete with the labelled ('hot') mRNA for the DNA. When, eventually, the unlabelled ('cold') mRNA was greatly in excess, little 'hot' mRNA could bind. Thus, you would see few double strands that were radioactively labelled.

On the other hand, suppose you mixed with the 'hot' mRNA unlabelled mRNA extracted from different tissues of a different organism (which we shall assume has a totally different set of mRNA molecules). Increasing the concentration of these will have little effect on the binding of the 'hot' mRNA to the DNA. This is because there will be no competition between 'hot' and 'cold' mRNA for the DNA, because the mRNA molecules have different sequences. The end result will be that all the double-stranded DNA–mRNA formed will be radioactively labelled.

competitive DNA–mRNA hybridization*

Figure 30 illustrates the principle of *competitive DNA–mRNA hybridization* as demonstrated in two hypothetical experiments.

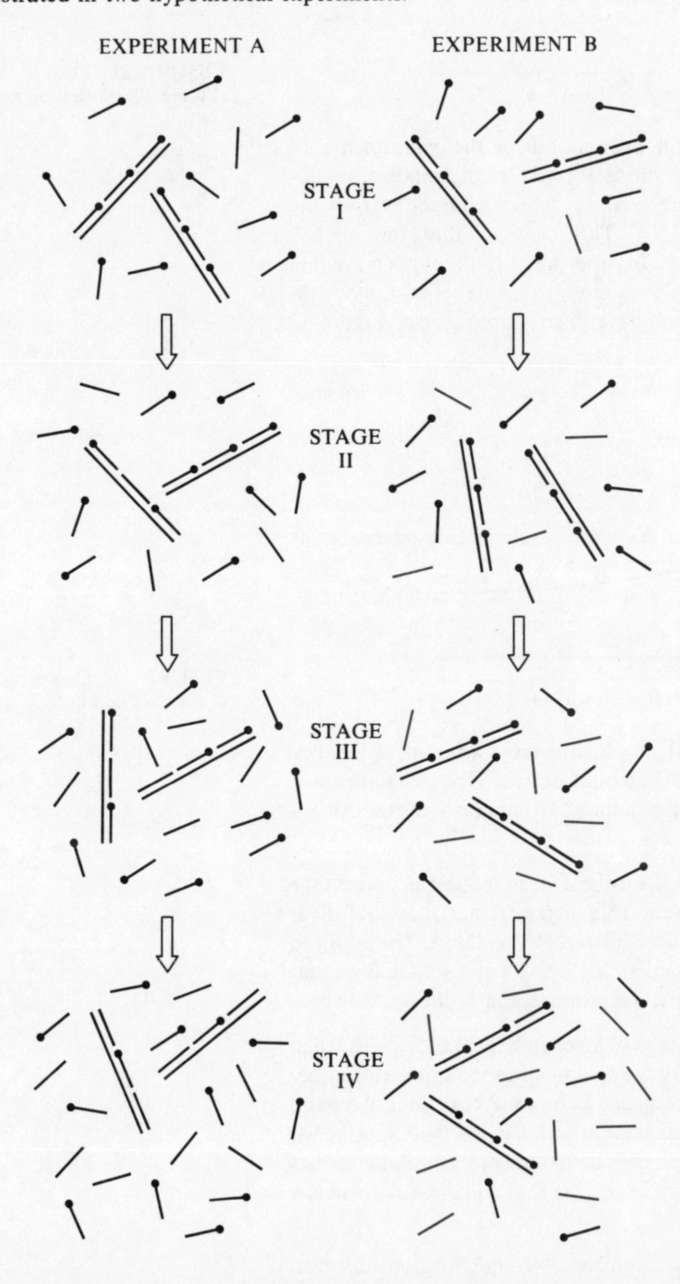

FIGURE 30 Competitive hybridization of DNA and mRNA—a model. In experiment A mRNA molecules (*short, black lines*) complementary to the DNA are hybridizing with the DNA and in experiment B uncomplementary mRNA molecules (*short, red lines*) are involved. This is a gross simplification: in practice there would be thousands of DNA molecules and magnitudes more mRNA.

In experiment A there is a constant amount of labelled mRNA (*short, black line with dot*); but the level of unlabelled mRNA (*short black line*) increases through stages I–IV. Both types of mRNA are complementary to the DNA. When concentrations of unlabelled mRNA are low (stage I), the only DNA–mRNA hybrids that form will be labelled. With increasing concentrations of unlabelled mRNA the proportion of hybrids with unlabelled mRNA will increase (see stage IV). Thus, the unlabelled mRNA is competing with the labelled mRNA for complementary sequences on the DNA.

In experiment B—where the types of mRNA are *not* complementary—labelled hybrids are formed initially as in experiment A. However, when the concentration of unlabelled mRNA (*red*) is increased, there is no reduction in the formation of labelled hybrids because there is no competition for the DNA from the unlabelled mRNA.

Experiments A and B would provide data that could be plotted as in Figure 31.

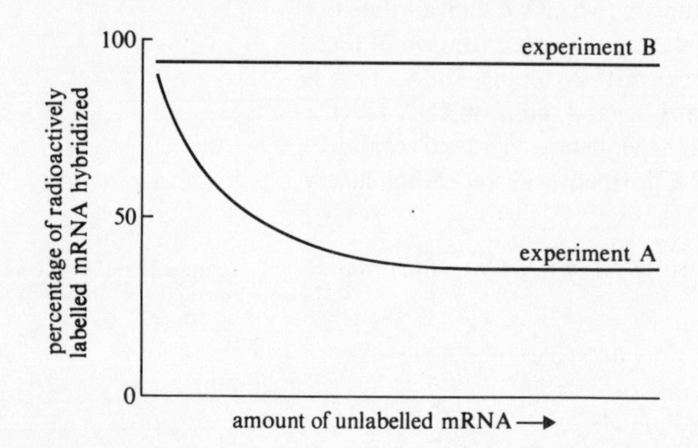

FIGURE 31 Data from the model (Figure 30) plotted on a graph.

The shape of the graph tells us what was happening during the experiments. In experiment A there were fewer labelled hybrids formed as the amount of unlabelled mRNA was increased. Thus, there *must* have been competition for the DNA between the two samples of mRNA. This suggests that the mRNA molecules must have very similar sequences. In experiment B there was no reduction in the number of labelled hybrids as the amount of unlabelled mRNA increased. Thus, these mRNA molecules must have different sequences. Figure 32

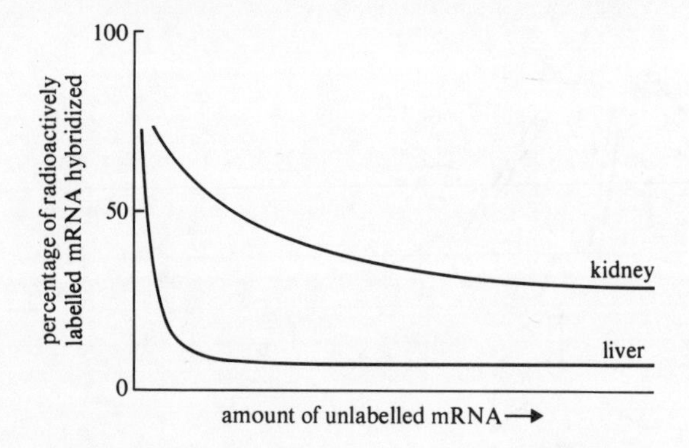

FIGURE 32 Competitive hybridization of rat liver and kidney mRNA.

shows the results from a real experiment in which a large amount of labelled mRNA from rat liver was added to a small amount of liver DNA to saturate it. This was done in the presence of increasing amounts of (i) unlabelled liver mRNA and (ii) unlabelled kidney mRNA.

Relatively small amounts of 'cold' liver mRNA cause a steep fall in the percentage of labelled DNA–mRNA hybrid molecules. This suggests that the 'cold' liver mRNA effectively competes with the 'hot' liver mRNA for the DNA. The addition of 'cold' kidney mRNA does not cause a similar fall. Thus, you could deduce that there *must* be differences in the types of mRNA molecules in kidney and liver.

However, there is a *slow* fall in the percentage of labelled hybrid molecules when kidney mRNA is added. This can only be due to some limited competition between the labelled liver mRNA and the unlabelled kidney mRNA. Thus, liver and kidney must have *some* similar mRNA sequences. In fact, despite their specialized roles in the body, both liver and kidney must contain a number of similar enzymes for basic metabolism. You would therefore expect to find some similar mRNA molecules coding for these proteins.

Experiments carried out in the 1960s using a similar DNA–mRNA hybridization technique claimed to show that there was a definite molecular basis for the differential transcription of genes in eukaryotes. This was because different sequences of mRNA were found in different tissues. The presence of these different sequences of mRNA was direct evidence for *differential gene transcription.* Unfortunately, as more was discovered about the nature of DNA, a major flaw was found in this work. You will recall from Section 6 that the DNA molecule itself has short repetitive sequences. By using a high concentration of mRNA compared

differential transcription of genes*

with that of DNA it was hoped that recombination of the DNA with itself would be limited. However, sequences of DNA may still rapidly recombine, which may influence the hybridization of DNA and mRNA.

The basic problem was that the technique involved the identification of a few specific sequences of DNA coding for mRNA among a large amount of additional DNA. Obviously, a different approach was needed. What was needed was relatively 'pure' DNA, which coded for a specific mRNA. This could then be used in hybridization studies with mRNA.

7.2 Hybridization using a DNA specific for mRNA

Two ways of getting 'pure' DNA that codes for a specific mRNA have been developed. One involves the extraction from cells of an mRNA that codes for a particular protein or set of proteins. A complementary sequence of DNA is then made on this mRNA template. In principle, this is quite simple. An enzyme extracted from a certain virus has the ability to transcribe its genetic information (which is stored as a sequence of RNA rather than DNA) in a reverse direction. If this enzyme, known as *reverse transcriptase*, is mixed *in vitro* with DNA and RNA nucleotides, DNA that is a complementary copy of the RNA will be synthesized. Such DNA is generally known as *cDNA* (*complementary DNA*).

reverse transcriptase

complementary DNA (cDNA)*

cDNA–mRNA hybridization*

This complementary DNA can be radioactively labelled and used in hybridization experiments with mRNA. Because the cDNA codes *only* for a specific mRNA we do not have the additional complication of searching for mRNA sequences on the DNA molecule with competitive hybridization techniques.

Increasing amounts of mRNA from different tissues can be mixed in turn with cDNA from one of the tissues. The relative rate at which labelled hybrids are formed, and how many are formed, will give an indication of how similar or dissimilar the mRNA molecules are. Alternatively, cDNA can be used to investigate changes in types of mRNA at different stages in the development of a tissue or organism.

Figure 33 shows the sorts of results one might obtain with cDNA prepared from RNA extracted from tissue X. This XcDNA was then hybridized over a period of time with increasing concentrations of mRNA from X and from another tissue Y.

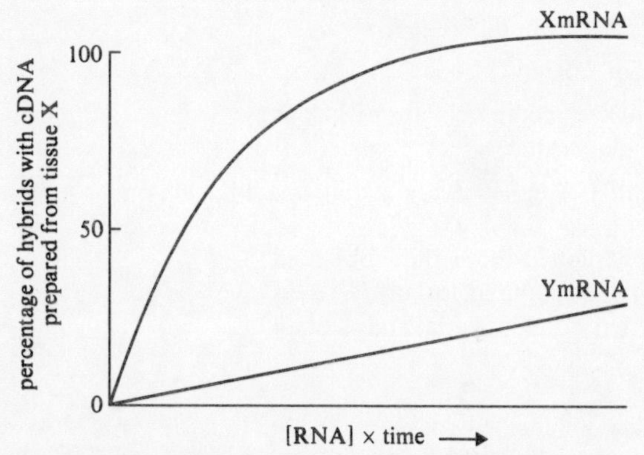

FIGURE 33 The hybridization of cDNA from tissue X with mRNA from tissues X and Y. (*Note* The square brackets around RNA on the *x* axis denote a concentration.)

You can see that when mRNA from X, the tissue that was used to prepare the cDNA, is hybridized with the cDNA, there is a rapid rise in the percentage of labelled hybrids formed. However, with mRNA from tissue Y the rate and total amount of hybrids formed are much lower than for XmRNA.

□ What can you tell about the similarity of the mRNA molecules from tissues X and Y in this experiment?

■ The mRNA from the two tissues, X and Y, cannot be very similar, otherwise they would have produced similar curves when hybridized with cDNA from tissue X.

You should realize that this technique using cDNA is useful only when it is possible to obtain a large quantity of 'pure' mRNA, that is in a system that is producing a large quantity of one particular protein. A large quantity of mRNA is needed for the synthesis of a large amount of cDNA, and the mRNA needs to be 'pure', otherwise the cDNA produced is complementary to a *range* of RNA molecules. Thus, it is not possible to use cDNA to investigate changes in mRNA in all developing organisms.

An alternative method of obtaining DNA that codes only for a particular mRNA is to extract DNA from cells and then separate out the single-copy fraction (scDNA). This codes in general for mRNA (Section 6.1). By repeated hybridization of this DNA with mRNA from a particular tissue, the DNA is gradually purified. In the end, DNA that codes almost exclusively for the mRNA from that tissue is obtained. This DNA is known as *messenger* or *mDNA* and in fact, is simply purified scDNA. It can be used in hybridization studies in a similar way to cDNA.

messenger DNA (mDNA)*

Because repeated hybridization allows the production of DNA that codes, in the main, only for the mRNA in that tissue or organism, this technique is useful in studies that compare changes in a whole range of mRNA molecules rather than in one particular type. In Section 8, you will learn how mDNA produced in this way has been used to investigate the differential transcription of genes in sea-urchins.

Summary of Section 7

DNA–mRNA hybridization involves either producing cDNA or purifying the DNA fraction that codes for mRNA to produce mDNA. The extent of hybridization between mRNA and either cDNA or mDNA indicates the degree of similarity between the mRNA molecules of different tissues or organisms.

Objectives and SAQs for Section 7

Now that you have completed this Section, you should be able to:

★ give reasons why *competitive* DNA–mRNA hybridization may not provide conclusive evidence for the differential transcription of genes.

★ state the principle of cDNA–mRNA and mDNA–mRNA hybridization.

To test your understanding of this Section, try the following SAQs.

SAQ 7 (*Objective 18*) Discuss whether each of the following statements provides good reason to doubt whether competitive DNA–mRNA hybridization offers conclusive evidence for the differential transcription of genes.

(a) It is not possible to distinguish mRNA molecules from different tissues.

(b) DNA codes for rRNA and tRNA as well as for mRNA.

(c) There are repeated sequences in the DNA that may combine with mRNA or with complementary sequences of DNA at a very fast rate.

(d) Some tissues have a number of very similar mRNA molecules.

SAQ 8 (*Objective 19*) Figure 34 shows a graph plotted from data obtained during an experiment in which 'pure' DNA (mDNA) coding for mRNA was hybridized with mRNA extracted from four different developmental stages of an organism (I–IV).

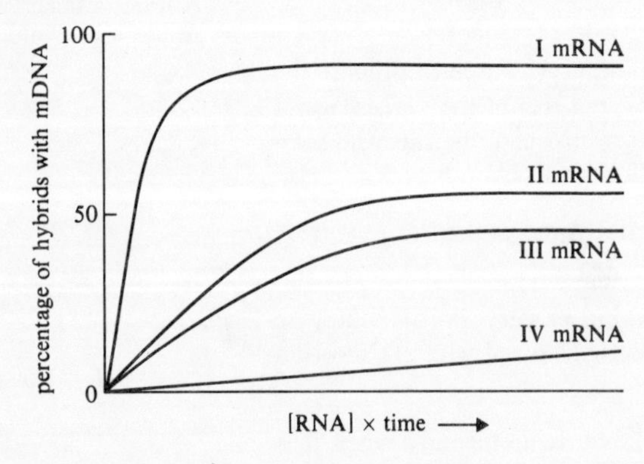

FIGURE 34 The results of the hybridization of mDNA and mRNA from four different developmental stages of an organism.

(a) Which of the developmental stages have similar kinds of mRNA?

(b) Which kind of mRNA has complementary sequences very similar to that of mDNA?

(c) At what developmental stage might the DNA that was used to obtain mDNA have been extracted?

8 | Experimental studies on gene expression

The experiments described in this Section use the DNA–mRNA hybridization technique, so it is important that you understand the basic principles as described in Section 7. If you have not already done so, you should work through the SAQs for Section 7, and if necessary re-read the Section before starting Section 8.

We shall consider first some experiments that have provided information about the possible role of the proteins associated with the DNA in controlling gene expression. Then the idea that development could be due to the control of DNA transcription will be reconsidered in the light of studies on sea-urchin development.

8.1 Chromatin proteins

As you know from Unit 5, Section 5.4, DNA is associated in the nuclei of eukaryotic cells with a number of proteins to form what is known as *chromatin*. These proteins are of distinct types: *histone proteins* and *non-histone proteins*. Experiments were carried out to investigate the importance of these chromatin proteins in controlling the transcription of genes using chromatin from calf thymus (an endocrine gland).

chromatin*
histone proteins*
non-histone proteins*

It is possible to separate these chromatin proteins from the DNA and to recombine them again. The amount of DNA that could be transcribed (measured by the amount of RNA produced *in vitro*) was then compared in intact chromatin, DNA on its own and DNA recombined with each of the two chromatin proteins. The results are shown in Table 4.

TABLE 4 DNA transcription from calf thymus chromatin and from the separated DNA and chromatin proteins

Template material for synthesis of RNA *in vitro*	Percentage of DNA transcribed
intact chromatin	5
DNA alone	40
DNA + histone protein	0
DNA + non-histone protein	14

These data show that 'naked' DNA transcribes more RNA than the intact chromatin. Thus, the chromatin protein must reduce, in some way, the transcription of the DNA. When DNA and the histone proteins are used as a template there is no transcription. When DNA and non-histone protein are used there is reduced transcription of the DNA.

These experiments suggest that the non-histone proteins have some controlling influence over the expression of the DNA. However, the combination of histone protein and DNA seems to stop all gene activity. You should note, though, that this study provides no information on the activity of *specific* genes.

8.2 The control of transcription during development—ovalbumin experiments

The experiments outlined in Section 8.1 demonstrate that the transcription of DNA is modified by the chromatin proteins. Is there any evidence that these proteins are involved in the differential and specific expression of the DNA during development? Studies on the production of ovalbumin, the 'egg-white' protein, have provided some information on this question.

Ovalbumin is produced in the oviduct* of mature hens or in immature hens that have been treated with the hormone, oestrogen.

* The oviduct is the tube through which the eggs of birds pass from the ovary to the outside. In the first section of the oviduct ovalbumin is produced and deposited around the egg. Further down, the oviduct produces the eggshell and mucus to assist egg-laying.

□ Knowing that the production of ovalbumin increases following treatment with oestrogen, what other molecules would you expect to find increasing in concentration in the cytoplasm of oviduct cells if genetic control of transcription were important?

■ mRNA molecules coding for ovalbumin.

The production of ovalbumin mRNA should precede the synthesis of ovalbumin. However, oestrogen may be acting by stimulating the translation of stored mRNA in the cytoplasm of the cells of the oviduct. In fact, we know that in an immature hen synthesis of mRNA is stimulated within 2 minutes of injecting oestrogen. This shows that the oestrogen is affecting the transcription of the DNA rather than the translation of stored mRNA.

When oestrogen enters the oviduct cell it first forms a complex with oestrogen-specific receptors. These are cytoplasmic proteins with a high affinity for oestrogen. Such a union is called a *hormone–receptor complex*. (You will learn more about this in Unit 16.) The complex will bind directly, but non-specifically, to oviduct cell DNA, but it is only when it binds to oviduct chromatin at specific binding sites that transcription of DNA to produce ovalbumin mRNA is evident. In fact, the oestrogen receptors will bind to any DNA—not just to DNA extracted from oviduct tissue. This indicates that the chromatin proteins are important in regulating the specific binding of the hormone–receptor complex to the DNA.

hormone–receptor complex

If the oviduct chromatin is treated so that the histone protein fraction is removed, one still finds specific binding of the hormone–receptor complex. This binding is retained even if the histones from a different tissue are recombined with the treated chromatin. It is only when the non-histone protein is removed that you find a reduction in the amount of binding of the hormone–receptor complex. Binding is increased when the non-histone proteins are replaced. If non-histone proteins from a *different* tissue are recombined with the treated chromatin there is no increase in binding.

In the reverse experiment, non-histone protein from oviduct cells can be recombined with histone and DNA from another tissue. When this is done the hormone–receptor complex binds to this 'new' DNA. Thus, like the calf thymus experiments, these experiments clearly show that the non-histone protein has an effect on the transcription of DNA.

The question of whether binding initiates the transcription of *specific* genes can be studied using the hybridization procedures described in Section 7. Complementary DNA coding for ovalbumin is made from ovalbumin mRNA. In this way it is possible to compare transcription from chromatin isolated from unstimulated oviducts and from oviducts treated with oestrogen. The results are shown in Table 5.

TABLE 5 The production of mRNA in unstimulated and oestrogen-treated oviducts

Hormonal state of oviduct	No. of molecules of ovalbumin mRNA per cell
unstimulated	0
after 4 days of treatment	20×10^3
after 9 days of treatment	44×10^3
after 18 days of treatment	48×10^3

You can see from these results that oestrogen treatment causes an increase in the synthesis of the mRNA that codes for ovalbumin.

From Sections 8.1 and 8.2 you will have gathered that non-histone proteins are thought to be important regulators of the transcription of DNA and that specific hormones can elicit the transcription of specific sequences of DNA *via* interaction with the non-histone proteins. In Section 8.3 we consider evidence that control over the translation of mRNA, rather than control over the transcription of DNA, may be important in some developmental systems.

8.3 Sea-urchin experiments: introduction

The experiments on sea-urchins described here were designed to show whether developmental changes occur via the control of the transcription of DNA or the translation of mRNA in multicellular organisms. Of course, you should have realized by now that trying to draw parallels between different organisms is fraught with difficulties. Any information gained from studies on sea-urchins tells us only about sea-urchins—at best such information can provide an hypothesis for testing with other multicellular organisms.

An advantage of using sea-urchins as experimental organisms is that it is relatively easy to obtain large quantities of eggs that develop synchronously and can then be analysed at various stages of development (Figure 35).

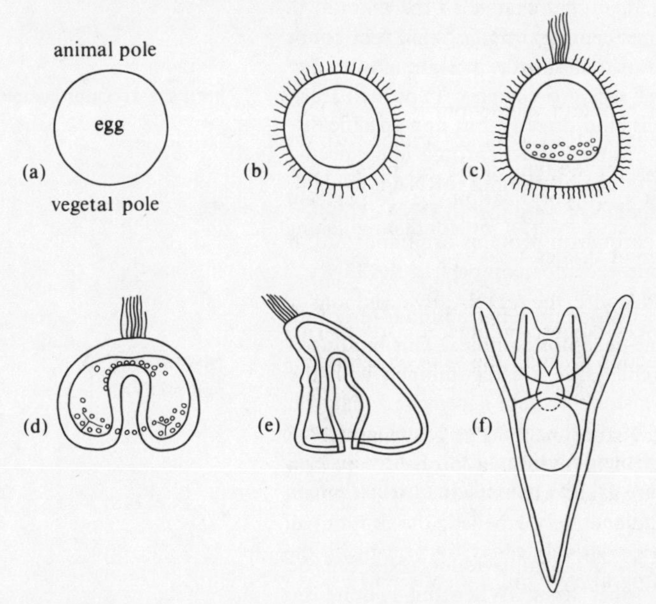

FIGURE 35 A schematic representation of sea-urchin development: (a) egg; (b and c) early and late blastulae; (d) gastrula; (e) prism; (f) young pluteus.

There is sufficient material at each developmental stage for a detailed analysis of sequences of mRNA to be made. If we found that there were *no differences* in the mRNA molecules present in the tissues of different developmental stages, *translational control* would be likely. Conversely, *changes* in mRNA molecules between the different stages would support the concept of *transcriptional control.*

8.3.1 The experiments

The technique used in these experiments is DNA–mRNA hybridization, which you read about in Section 7. Recall that it is possible to 'purify' DNA so as to obtain a DNA that codes, almost exclusively, for mRNA.

The technique involves the extraction of DNA from sea-urchin sperm and the isolation of the DNA sequences coding for mRNA. Taking DNA from sperm ensures that all the mRNA produced during early development would be coded for by the DNA. If DNA was taken from one particular stage of development there would be a possibility, albeit very remote, that some DNA had been lost during development. In fact, we are reasonably sure that all sea-urchin cells have the same genetic information, so you could use DNA from any stage of development. Because the technique involves the isolation of specific sequences of mRNA, there is no confusion with other RNA molecules. In outline, this extraction involves denaturing the DNA duplex by heating it and then letting it hybridize slowly on cooling. The DNA sequences that take the longest time to hybridize are those that have a unique sequence. As you know from Section 6, this DNA is known as *s*ingle *c*opy DNA (or scDNA). It is this scDNA that codes for mRNA.

DNA–mRNA hybridization experiments can be conducted with this scDNA using mRNA molecules from the stages and/or tissues of interest. To start with, the affinity of gastrula mRNA for this scDNA is determined. The results of hybridization of scDNA and gastrula mRNA show that only about 1 per cent of the DNA is hybridized with the mRNA.

□ What can you deduce from this result?

■ Remember that because the scDNA codes for *all* the mRNA in sea-urchins (because it was extracted from sea-urchin sperm), of the possible mRNA molecules that the DNA codes for, only a very small proportion can be present at any one time in the cells of the gastrula stage.

We shall consider later the significance of the fact that very little of the DNA coding for mRNA is expressed in the gastrula cells. However, the small amount of hybridization does pose problems when mRNA from other developmental stages is compared with gastrula mRNA. As very little of the mRNA is complementary to the DNA, small changes in the type of mRNA present at different times are likely to be missed by this technique. For example, if there was a 20 per cent change in the number of mRNA molecules present (i.e. a 20 per cent increase in mRNA) this would be reflected by only a 0.2 per cent change in the amount of mRNA–DNA hybridization (20 per cent of 1 per cent)—and this might well not be detected, given the inevitable experimental errors that occur with this technique.

The trick that was used to get around this problem was to purify the scDNA fraction to such a point that about 60 per cent of DNA would hybridize with the mRNA. Now, if there were a 20 per cent change in the type of mRNA present one would see a 12 per cent change in the amount of hybridization (i.e. 20 per cent of 60 = 12 per cent). This technique provides a very powerful tool for distinguishing between relative amounts of mRNA in different tissues.

By repeatedly hybridizing the gastrula mRNA with the scDNA, two fractions of DNA could be obtained, one with sequences corresponding only to gastrula mRNA—*gastrula mDNA*—and a second fraction with all the scDNA sequences that did not code for gastrula mRNA—*null mDNA*.

gastrula mDNA*
null mDNA*

The null mDNA can be used as a control in experiments. The gastrula mDNA (the DNA that codes for gastrula mRNA) can be used as a 'probe' to see how similar or dissimilar the mRNA sequences are at different stages of development or in different tissues. If this mDNA is radioactively labelled, the amount of DNA–mRNA hybridization can be detected relatively easily by measuring the radioactivity of hybrids isolated by chromatography (Unit 4, Section 4.1).

We shall now consider a number of experiments on sea-urchins and then try to draw some general conclusions.

EXPERIMENT 1

The reaction of gastrula mDNA and null mDNA with mRNA from adult intestinal cells was investigated. As a control experiment, the reaction of gastrula mRNA with the two DNA fractions was also looked at (results in Figure 36).

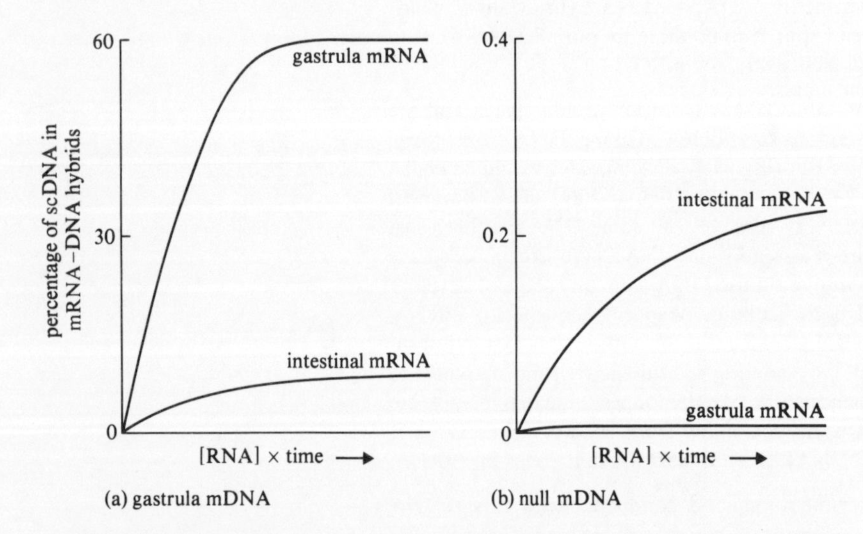

FIGURE 36 The reaction of (a) gastrula mDNA and (b) null mDNA with mRNA from the gastrula stage and from adult intestinal cells. (Do not worry about the vastly differing scales between the gastrular mDNA and the null mDNA experiments. These simply reflect different degrees of hybridization when these different preparations are used. The important points to note in this Figure and in Figures 37, 38 and 40 are the relative differences in the curves.)

□ What can you deduce from Figure 36 about the amount of gastrula mRNA present in the adult tissue?

■ Figure 36a shows that only a very small fraction of the mRNA present in the gastrula is present in the adult intestinal tissue because no more than about 8 per cent of the mRNA from this tissue hybridizes with the gastrula mDNA. From Figure 36b you can see clearly that in the adult tissue the null mDNA must code for mRNA molecules that are *not* present in the gastrula. This is because there is hybridization only between the null mDNA and the intestinal mRNA.

EXPERIMENT 2

This looked at the reaction of the two DNA fractions with mRNA from mature *oocytes* removed from sea-urchin ovary. The oocytes will grow to become the mature eggs released at spawning. Again, the reaction with gastrula mRNA was investigated. Figure 37 shows the results of the reaction of the oocyte mRNA with two fractions from the gastrula.

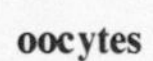

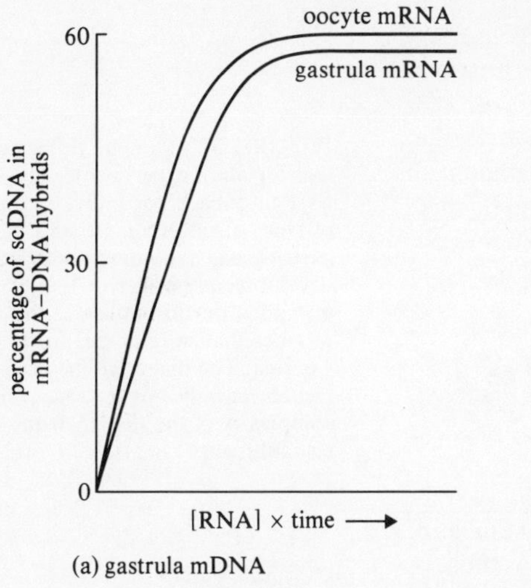

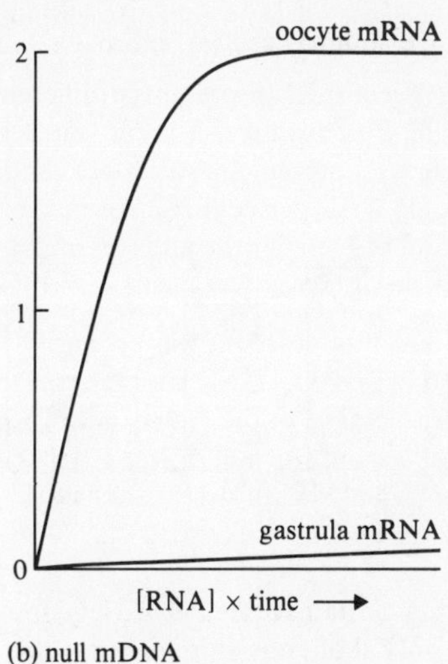

FIGURE 37 The reaction of (a) gastrula mDNA and (b) null mDNA with gastrula and oocyte mRNA.

□ What can you deduce from this Figure?

■ From Figure 37a you can see there is a great similarity between the oocyte mRNA and gastrula mRNA. Thus, all the mRNA in the gastrula is already present in the oocyte. On the other hand, from Figure 37b, it is clear that the oocyte also has some mRNA of a type that is not present in the gastrula because there is hybridization with the null mDNA, which we know has *no* sequences for gastrula mRNA.

Because the curves for gastrula mRNA and oocyte mRNA are so similar, we must assume that the genes necessary for gastrulation have been *transcribed* during the development of the egg and *before* fertilization. You might have expected that the unfertilized egg would have *less* mRNA than the more complicated gastrula stage. However, we see that all the mRNA in the gastrula is already present in the oocyte. The oocyte thus appears to contain an amount of *stored mRNA* which is available for translation when the egg develops after fertilization.

stored mRNA

Now, consider some experiments that look at the reaction of gastrula mDNA with mRNA molecules from other embryonic stages.

EXPERIMENT 3

The mRNA molecules from the earlier blastula stage and a later larval stage, known as the *pluteus*, were investigated. Using the same technique as in the previous experiments, the data shown in Figure 38 were obtained.

pluteus*

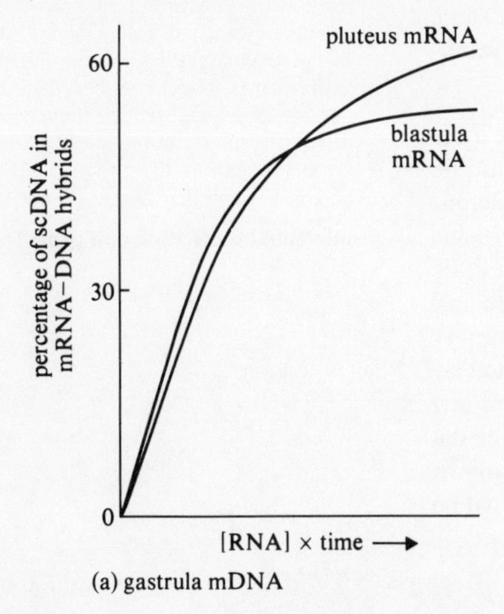

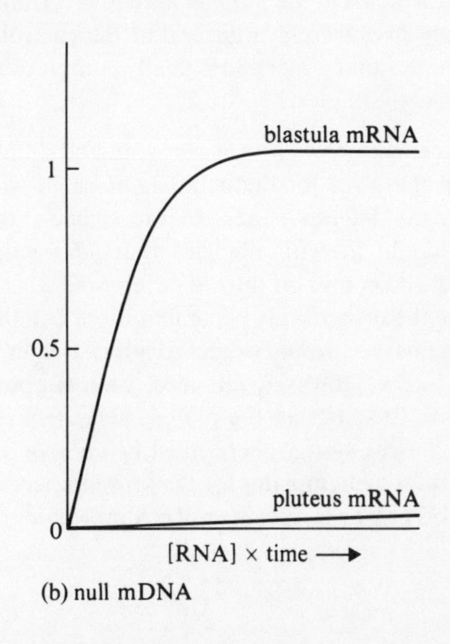

FIGURE 38 The reaction of (a) gastrula mDNA and (b) null mDNA with mRNA from the blastula and pluteus stages.

□ What can you deduce from these graphs?

■ From Figure 38a it is clear that a high proportion of the mRNA found in the gastrula is present in both the blastula and pluteus stages. However, Figure 38b suggests that the blastula contains mRNA molecules that are *not* present in the pluteus. At the pluteus stage, it appears that very few extra types of mRNA are found because there is no reaction with null mDNA.

The results of all three experiments are summarized in Figure 39.

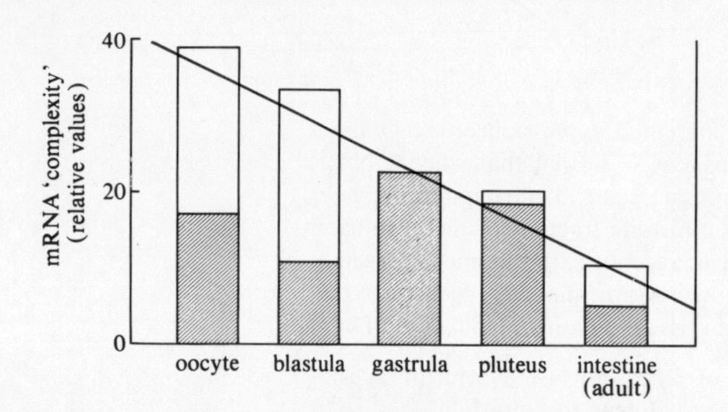

FIGURE 39 Summary of sea-urchin hybridization experiments. The shaded portion of each bar indicates the number of types of mRNA molecules that the gastrula has in common with other developmental stages or tissues. The unshaded portion indicates the number of types that were absent from the gastrula. The diagonal line across the histogram indicates the reduction in the complexity of the mRNA from oocyte to adult cells.

You can see that, although the blastula contains a greater variety and number of mRNA molecules than the gastrula, only about one-third are the same as those in the gastrula.

□ What can you say about the adult intestinal mRNA compared with the oocyte mRNA?

■ There is a great reduction in the total number and variety of mRNA molecules, both in terms of those held in common with the gastrula and those that are absent from the gastrula.

If you compare the pluteus with the gastrula you can see that the total amount of mRNA has changed very little between them. What is surprising is that there are relatively few 'new' types of mRNA molecules. Most of the mRNA in the pluteus is of the same type as that in the gastrula.

We can make three deductions from Figure 39.

1 The differences in the mRNA present in the gastrula compared with the other larval stages and with adult tissues must mean that large numbers of different genes are being transcribed at different stages of development (i.e. about half of the oocyte and intestinal mRNA has nothing in common with gastrula mRNA).

2 During development, the number of different mRNA molecules transcribed from the DNA is considerably reduced; that is, the 'complexity' of the mRNA is reduced. (The height of the histogram falls from a value of about 40 for the oocyte to about 10 for the intestinal cells.)

3 As there are few additional types of mRNA in the pluteus compared with the gastrula stage and the pluteus is a more complex structure, then *either* the mRNA molecules necessary to code for the proteins in the pluteus were in fact transcribed in the gastrula *or* the proteins themselves were synthesized at the gastrula stage but 'stored' in some way. (This assumes that a morphologically complex development stage is also more biochemically complex.)

What can we conclude from these deductions? The answer would be simple if the experiments had provided us with the basis for deduction 1 *alone*. It would be clear that, at least in sea-urchins, many different genes are transcribed at different times during development. This would fit with the idea that *differential gene transcription* (i.e. genes being switched on and off during development) is a basis for cellular differentiation. Because the organism is becoming more complex and thus needs a different range of proteins, you would expect to find a greater variety of mRNA in later developmental stages. However, this idea is *not* supported by deductions 2 and 3. Compared with the gastrula the pluteus stage has very few additional types of mRNA. Thus, in this system there must be control over the translation of the mRNA that provides the templates for the proteins necessary to build the more complex stages. In other words, specific mRNA molecules must be

differential transcription of genes*

translated at particular times. Alternatively, the proteins necessary for the development of the pluteus must have been made at an earlier developmental stage. This alternative idea is possible. Consider, for instance, why the 'simple' blastula has about 50 per cent more mRNA than the 'complex' pluteus? Possibly, this 'extra' mRNA codes for proteins present in the pluteus. The arguments here are involved, but it is clear that there is more to development in sea-urchins than controlled transcription of DNA, followed by translation of mRNA to produce an ordered sequence of proteins. This study has shown that there must also be control over the translation of mRNA.

Summary of Section 8

Non-histone proteins have a role in the differential expression of the DNA in some tissues. Experiments with calf thymus DNA showed that when DNA is associated with chromatin, limited transcription occurs. Transcription increases when DNA is associated with just the non-histone fraction of the chromatin proteins. Non-histone proteins bind to specific receptors; this stimulates ovalbumin production in the chick oviduct. Experiments show that oestrogen, the signal for ovalbumin production, acts by binding to the chromatin of oviduct cell DNA.

Finally, experiments using the technique of DNA–mRNA hybridization with sea-urchin material make it clear that differential gene transcription does occur because different types of mRNA are produced during sea-urchin development. However, this is not a full explanation of the control of development because the *more* complex larval stages have *fewer* kinds of mRNA. Any theory proposed about the control of sea-urchin development must be able to account for control of the translation of RNA.

Objectives and SAQs for Section 8

Now that you have completed this Section, you should be able to:

★ outline an experiment that shows that non-histone proteins may affect the transcription of genes.

★ describe how DNA–mRNA hybridization techniques have been used to investigate sea-urchin development and interpret the observations.

★ recognize that the process of development may involve the controlled expression of mRNA and proteins produced at earlier developmental stages rather than control over the transcription of the DNA.

To test your understanding of this Section, try the following SAQs.

SAQ 9 (*Objective 20*) Write a paragraph or so on the possible role of chromatin protein in cellular differentiation.

SAQ 10 (*Objective 21*) Figure 40 shows two graphs plotted from DNA–mRNA hybridization data. One is from the hybridization of mRNA and gastrula mDNA; the other is from the hybridization of mRNA and null mDNA.

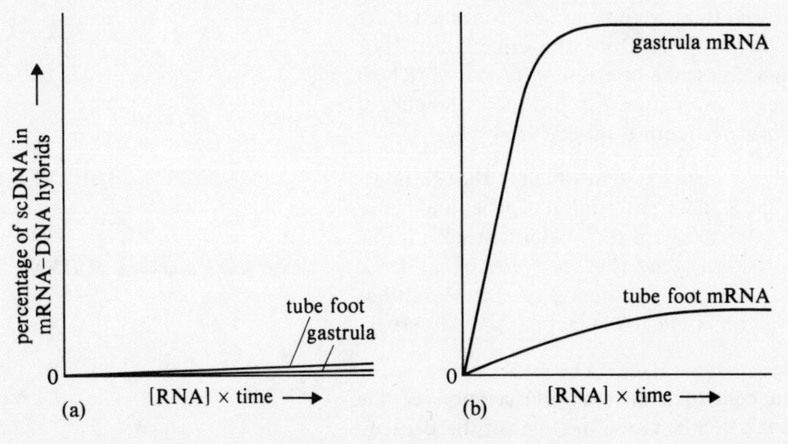

FIGURE 40 The reaction of gastrula mDNA and null mDNA with mRNA from the gastrula stage and adult tube foot cells.

(a) Which graph was derived from the hybridization of mRNA and DNA coding for gastrula mRNA? Explain the reason for your answer.

(b) Why are the curves for the hybridization of mRNA from tube foot cells different in Figures 40a and 40b.

(c) What are the implications of the differences between the curves for tube foot mRNA? (You should note that tube feet are present only after metamorphosis begins.)

SAQ 11 (*Objective 22*) What specific points of experimental evidence from the sea-urchin studies suggest that differential gene transcription alone cannot account for the way in which development and differentiation occur in sea-urchins?

9 The control of cellular differentiation

In this short Section we consider how the models for differential gene expression you met in Sections 5 and 6 can be applied to developing systems. It is important now that you refresh your mind on the ideas in Sections 5 and 6 by reading the relevant summaries.

The ovalbumin and sea-urchin experiments (Section 8) provide good experimental evidence for differential gene transcription. Although we cannot cover other systems in this Course, you should be aware that *direct* evidence for differential gene transcription has been found in other developing organisms. A good example is the development of lens tissue in the chick, and this is described in the TV programme, *Differential Gene Transcription.* However, at present (1980), the range of organisms providing direct evidence for the differential transcription of genes is fairly limited.

With this in mind, you may feel that any discussion of how differential gene transcription could be controlled is, to some extent, academic! However, it is possible to consider the Jacob–Monod and Britten–Davidson models (discussed in Sections 5 and 6) and see how these could fit in with the control of gene expression in developing systems.

First, we must consider an interesting feature of the nature of mRNA.

9.1 DNA to mRNA

Let us consider the concept that a precise sequence of DNA (a gene) codes for a specific sequence of mRNA, which acts as a template for the protein. A Britten–Davidson type of control system assumes that control is exercised over the transcription of a sequence of nucleotides which make up a length of mRNA.

However, research (1977/78) showed that control over transcription alone cannot explain how the cell produces a functioning mRNA molecule. It is possible to sequence RNA molecules in a precise way so that the order of bases is known. (The experimental details are very complex and will not be described here.) Similarly, one can obtain the sequence of bases on the DNA that codes for the RNA molecule. When this was done for a number of transfer RNA molecules in yeast the result was surprising. The DNA sequence had two regions of bases that were complementary to the tRNA bases but these were separated from each other by a sequence of DNA that did *not* code for tRNA. Once transcribed, the tRNA sequence would not be able to function unless the intervening sequence (RNA) was removed and the two tRNA sequences were joined to produce a functional tRNA molecule. In yeast cells these processes require specific enzymes.

At the time this was considered a rather unusual system which, although interesting, did not have much relevance to the actual transcription of genes and hence the production of proteins. However, later work on the ovalbumin gene in the chick has provided evidence for a structural gene that has *intervening DNA sequences* which do not code for RNA. With the ovalbumin gene, research has shown that the DNA that codes for ovalbumin mRNA in fact has *seven* intervening DNA sequences!

intervening sequences of DNA*

So, although there must be precise control over the transcription of the ovalbumin-specific mRNA, and this may well be along the lines of the Britten–Davidson model, there must be mechanisms within the cytoplasm to 'remove' the

tailoring and splicing of mRNA*

sequences of bases within the molecule that are not part of the functional mRNA and then splice the remaining mRNA pieces together so that translation of the mRNA is possible.

The analysis of gene structure is very complex and time-consuming so that only a limited number of genes have been studied so far. It would seem that the presence of intervening sequences of DNA must be a fairly general phenomenon in eukaryotic genes—they have been identified in β-globin genes, immunoglobulin genes, insulin genes and ribosomal genes in some species. However, it is clear that not all genes contain such sequences; for instance, they are absent from the genes for histone proteins.

The reason for discussing this work, apart from its scientific interest, has been to show you that the basic concept of DNA → mRNA → protein is an oversimplification. Because there is modification of the RNA after transcription it is important that any model control system for differential gene expression takes into account all the features of the system under study rather than just the basic ones.

9.2 The complexity of DNA and RNA—is the Jacob–Monod model valid for cellular differentiation?

Let us draw some strands together.

1 In Unit 5 and the TV programme, *Differential Gene Expression*, you learned about the nature of chromatin. In this Unit some of the complexities of DNA and RNA in eukaryotes have been mentioned. It should be clear to you that eukaryotic chromatin is thus far more complex than the genetic material in prokaryotes.

2 You also know that DNA codes for rRNA and tRNA as well as mRNA (Section 6).

3 From Section 9.1, you will have realized that (a) RNA may be modified in the cytoplasm *after* transcription and (b) in some organisms at least the DNA does code *directly* for a precise complementary sequence of mRNA.

CHROMATIN

The fact that eukaryotic DNA is closely associated with histones and other proteins complicates the application of the Jacob–Monod model. However, because the DNA is, as far as we know, on the outside of the chromatin proteins it is possible that the 'control' points on the molecule would be accessible. In Section 8 we mentioned the importance of non-histones in the activity of specific genes and rather ignored any other features of the chromatin proteins. It seemed clear from the experiments mentioned that the histones had an inhibitory role. However, there is good evidence now (1980) that different types of histones are involved at different times during differentiation in some cells. For example, some histones could react with the DNA so as to expose certain portions of DNA at particular times. However, there is always the chicken and egg argument—do chromatin proteins change their structure to allow differential transcription of the DNA or are those changes a *result* of transcription?

DNA THAT DOES NOT CODE FOR mRNA

There is no reason why this should affect our models for control. All it means is that the DNA sequence coding for mRNA will be spaced out along the chromatin. It is important that, as well as controlling the production of mRNA, the cell has a way of controlling the production of tRNA and rRNA. This will obviously not need to be so precise because the synthesis of proteins will generally require the whole range of tRNA molecules plus sufficient ribosomes, rather than one or two specific tRNA molecules in large quantities.

MODIFICATION OF mRNA AFTER TRANSCRIPTION

Although control over the production of mRNA could operate along the lines of the Jacob–Monod model, there must be some precise mechanism for tailoring the mRNA after transcription to cut out intervening sequences. However, this may not need any specific control because there may be sequences of bases that enzymes recognize and cause a break in the mRNA molecule at the correct point. One fact we did not mention earlier is that far more mRNA is produced in the nucleus than ever reaches the cytoplasm. Thus there must be selection in some way of the appropriate mRNA molecules by factors in the nucleus. Comparison of mRNA sequences from nuclear and cytoplasmic fractions shows that the sequences in the nucleus (immediately after transcription) are larger than those in the cytoplasm. The significance of this is, as yet, unknown. It is, however, the difference in the amount of mRNA between nucleus and cytoplasm and the difference in size that indicates that there may be another level of control of the expression of DNA over and above the regulation of transcription of the DNA along the lines of the Jacob–Monod model.

From this you will realize that the Jacob–Monod model could work for eukaryotic organisms. There are, however, a number of points that the model cannot cope with, given the current ideas on the relationship between DNA and mRNA. In particular, the fact that much more mRNA is transcribed than ever reaches the cytoplasm is very difficult to explain in these terms. It is obvious, therefore, that control over the differential transcription of genes is likely to be more complex than the basic Jacob–Monod model. We hinted at this when we discussed Britten's and Davidson's model. You should now be able to see that even their model cannot account for the variety of control systems needed to produce functional mRNA in the cell at a particular time. As a final note, we have not even started to consider how the control of proteins, mentioned in the work on sea-urchins (Section 8), may operate!

Summary of Section 9

1 Not all DNA codes for mRNA.

2 The large amount of DNA transcribed but the presence of only a small amount of mRNA in the cytoplasm is difficult to explain.

3 mRNA is 'tailored' and 'spliced' in the cytoplasm.

4 Control over which particular mRNA molecules reach the cytoplasm must be exercised in the nucleus.

5 The models for control over the production of mRNA are generally correct but control exercised *after* the transcription of the DNA may well be crucial.

Objectives and SAQs for Section 9

Now that you have completed this Section, you should be able to:

★ give reasons why the concept that DNA codes for a precise sequence of mRNA may be incorrect.

★ outline how current ideas on DNA and RNA in eukaryotic cells fit the Jacob–Monod model.

To test your understanding of this Section, try the following SAQs.

SAQ 12 (*Objective 23*) Does the following experimental observation provide any evidence that DNA may not code for a precise sequence of mRNA? Give reasons for your answer.

'mRNA extracted from the nuclei of chick cells that synthesize ovalbumin could not be translated into ovalbumin in *in vitro* experiments. However, mRNA taken from the cytoplasm of those cells was able to produce ovalbumin.'

SAQ 13 (*Objective 24*) Write one or two sentences outlining how each of the following affects, in principle, the application of the Jacob–Monod model to eukaryotic organisms:

(a) chromatin

(b) DNA that does not code for mRNA

(c) the modification of mRNA after transcription.

10 Postscript

Towards the end of these Units we have inevitably taken a very molecular approach to the study of cellular differentiation. Once we established that the differential transcription of genes was a plausible explanation of cellular differentiation we were led to question how transcription could be controlled. The complexities of DNA and chromatin structure have to be taken into account. You should realize that making our approach to this topic in ever more detail does not necessarily mean that we are any nearer an 'explanation' of cellular differentiation. However, we may well be asking more relevant questions.

Where does cellular differentiation fit into the whole scheme of development? Even if we could list all the changes that occur to all the cell types in any one organism, would this in itself allow us fully to describe or predict the form that the organism will take? Consider, as an example, a human hand or foot. Both contain skin, bone, muscle, and so on, and are made up from identical types of cell. Yet they are clearly different from each other and occur at different places in the body. So knowing the changes that occur to individual cells as they undergo differentiation is not enough; knowing how these cells interact is also essential to any understanding of development. It is these problems concerning the changes, signals and genetic information in developing systems that are covered in the next two Units.

Objectives for Units 12 and 13

Now that you have completed this Unit, you should be able to:

1 Define and use, recognize definitions and applications of the terms marked by an asterisk in Table A.

2 Outline the basic assumptions underlying the study of cellular differentiation.

3 Give reasons why it is valid to consider that genetic material is constant between the cells of most multicellular eukaryotic organisms. (*SAQ 1*)

4 Describe how plant cell-culture and nuclear transplantation studies have provided good evidence for totipotency in fully differentiated cells. (*SAQ 2*)

5 Give examples of organisms in which totipotency cannot occur. (*SAQ 2*)

6 Outline the main factors that could maintain cells in a fully differentiated state. (*Tested by ITQ*)

7 Describe the contribution of cell-fusion experiments to our understanding of the interactions of nucleus and cytoplasm. (*SAQ 3*)

8 Outline the experiments that provide the evidence for the control of the translation of cytoplasmic mRNA in *Acetabularia*. (*SAQ 3*)

9 Give one example of biochemical changes associated with a developmental process.

10 Outline three hypotheses that could account for differentiation. (*SAQ 4*)

11 Outline the problems of designing experiments to test the relative roles of protein synthesis and degradation in developing organisms.

12 Outline the Jacob–Monod model of the control of enzyme induction in bacteria. (*SAQ 5*)

13 Give examples of enzyme induction in bacteria which may involve control at the level of translation rather than at the level of transcription.

14 Outline the ways in which eukaryotic DNA is more complex than prokaryotic DNA. (*SAQ 6*)

15 State the principle of DNA–DNA hybridization and interpret observations. (*SAQ 6*)

16 Give reasons why a more complex method of control than the Jacob–Monod model may be needed in eukaryotic organisms. (*SAQ 6*)

17 Outline the Britten–Davidson model for the control of gene expression in eukaryotic organisms. (*SAQ 6*)

18 Give reasons why competitive DNA–mRNA hybridization may not provide conclusive evidence for differential gene transcription. (*SAQ 7*)

19 State the principle of cDNA–mRNA and mDNA–mRNA hybridization. (*SAQ 8*)

20 Outline an experiment that shows that non-histone proteins may affect gene transcription. (*SAQ 9*)

21 Describe how DNA–mRNA hybridization techniques have been used to investigate sea-urchin development and interpret observations. (*SAQ 10*)

22 From sea-urchin studies, recognize that the process of development may involve the controlled expression of mRNA and proteins produced at earlier developmental stages rather than control over the transcription of the DNA. (*SAQ 11*)

23 Give reasons why the concept that DNA codes for a precise sequence of mRNA may be incorrect. (*SAQ 12*)

24 Outline how our current ideas on DNA and RNA in eukaryotic cells fit in with the Jacob–Monod model. (*SAQ 13*)

Pre-Unit test answers and comments

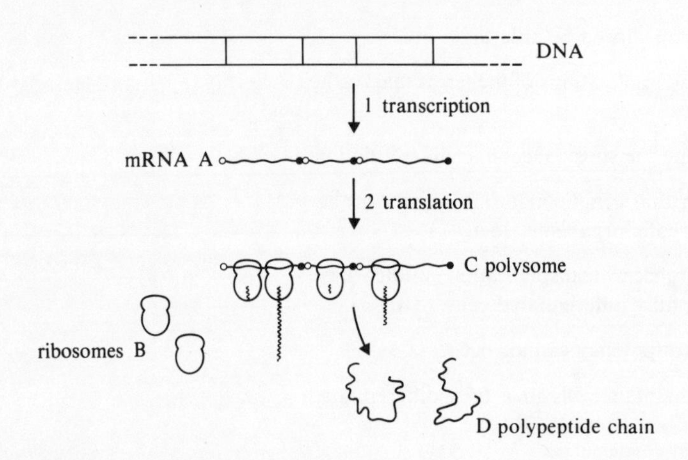

FIGURE 41 (Figure 1 completed).

(a) See Figure 41 (completed Figure 1).

TRANSCRIPTION

A gene is copied to give an mRNA molecule. That is, the sequence of bases of one strand of one region of the DNA double helix is copied to give a complementary strand of RNA. This requires nucleotides with the four bases adenine, uracil, cytosine, and guanine. The copying is catalysed by an enzyme that polymerizes the nucleotide so that a linear molecule of mRNA is built up.

TRANSLATION

A ribosome attaches to the mRNA. A tRNA molecule with its associated amino acid binds to the ribosome and to the mRNA. Another tRNA molecule binds to the adjacent site and a peptide bond is formed between the two amino acids. The ribosome now moves and enables a new amino acid–tRNA complex to attach to the mRNA and ribosome. In this way a polypeptide chain is formed, which then folds to form the protein.

(b) Nucleotides are needed at stage 1, which is the transcription of DNA.

(c) Transfer RNA molecules joined to amino acids by activated enzymes, ribosomes and messenger RNA are needed for stage 2, which is the translation of mRNA.

(d) See Figure 41 (completed Figure 1).

ITQ answers and comments

ITQ 1 If the nucleus is damaged, this may well affect the genetic information it contains. The higher failure rate with older nuclei could result from the greater likelihood of nuclear damage rather than the lack of genetic information. Thus scheme B may well be more likely than scheme A.

ITQ 2 You could inject a similar volume of cytoplasm from the donor cell or, alternatively, an *inactivated* nucleus into the frog egg. Any development from such a control experiment would raise serious doubts as to whether the results were artefacts due to the experimental treatment.

ITQ 3 From point 1, contact between the cells seems to be crucial in maintaining the differentiated state. We also know that the particular environment the cells find themselves in is important because altering the chemical composition of the medium can affect the way the cells respond.

From point 2, it is clear that the cytoplasm of the egg can, in some way, act on the nucleus of a fully differentiated cell to allow the expression of its genetic information. It follows that the cytoplasm of the differentiated cell may act to prevent the expression of some of the genetic information within the nucleus.

ITQ 4 The obvious answer must be no! DNA *synthesis* has nothing to do with the differential *transcription* of genes. However, it is important to some extent in that it shows that some factors in the cytoplasm are interacting with the DNA. In the HeLa–macrophage and HeLa–lymphocyte fusions the synthesis of DNA is stimulated when macrophage and lymphocyte nuclei are introduced into HeLa cytoplasm.

ITQ 5 The results of the first experiment suggest that cytoplasmic factors may affect cap morphology, so the reason for a hybrid cap may be that there is a combination of nuclear and cytoplasmic factors. A normal cap is formed only when both the nucleus and cytoplasm have the same 'information'.

ITQ 6 1 The induction of certain enzymes occurs when substances that are substrates of those enzymes, or analogues of substrates, are present in the cells. Induction begins within minutes.

2 Induction involves the synthesis of new polypeptide chains, not just the modification of existing ones.

3 Bacterial mRNA has a short half-life.

4 The control of the synthesis of polypeptides could be exercised through the control of the rates of transcription or of translation.

5 The Jacob–Monod model predicts that:

(a) control is via the regulation of transcription;

(b) the structural genes for enzymes that are induced or repressed together are adjacent to each other on the DNA;

(c) each set of structural genes has a specific operator region adjacent to it;

(d) each operon is regulated by a specific regulator gene;

(e) each regulator gene contains information for a specific repressor;

(f) the repressor binds to its specific operator region, thus preventing RNA polymerase from transcribing the structural genes.

(g) the binding of the repressor to the operator region is prevented by the inducer;

(h) in repression, the binding of the repressor to the operator region is facilitated by the presence of a small co-repressor molecule.

ITQ 7 Some enzymes—constitutive enzymes—are made continually because their amounts are not influenced by the level of substrates in the cells.

SAQ answers and comments

SAQ 1 (a) This suggests that the genetic material is constant, but there is no way of knowing for certain that the genetic information (i.e. the DNA) is identical.

(b) Because banding is due to the ordered arrangement of DNA, this statement gives more weight to the idea that genetic information is constant between cells, but again it is not conclusive.

(c) Again, this *suggests* only that the genetic information is constant.

(d) This is the best statement because it takes into account a number of parameters relating to the constancy of genetic information.

You should realize that evidence for this concept is to a large extent circumstantial and is derived from a wide range of studies. There is no one *definitive* piece of evidence. The best work we have is the study on insect chromosomes. Refer back to p. 7 if you were not able to answer this question.

SAQ 2 (a) True. As long as the cells contain the same genetic information they will be totipotent. The exceptions will be fully differentiated cells that do not have chromosomes (e.g. mammalian red blood cells).

(b) False. Contact between cells would appear to be a crucial factor because when tissues are broken up the separate cells can differentiate into a whole plant.

(c) False—or rather not strictly true. Work with plant cell cultures has provided evidence just as good as that from nuclear transplantation studies.

(d) False. The genetic information may be affected but is unlikely to be lost. The most plausible explanation is the occurrence of chromosomal abnormalities that affect the ability of the host cell to divide.

(e) True. (See the answer to ITQ 3).

(f) False. Gall midges are like *Parascaris* sp. in that they lose chromosomal material during development.

SAQ 3 True. The hen erythrocyte experiments show conclusively that the cytoplasm of a HeLa cell can stimulate RNA synthesis in the nucleus of the hen erythrocyte.

(b) False. Cytoplasmic signals are not species-specific, but you can see from Table 2 that rabbit macrophages do not produce DNA so you would not expect the macrophage to stimulate the synthesis of DNA in the hen erythrocyte nucleus. It is unlikely then that cytoplasm could direct DNA synthesis in the nucleus of the hen erythrocyte cell.

(c) False. The experiments on p. 15 show that the result of such a procedure is to produce a *hybrid* cap.

(d) True in principle. Ribonuclease will break down any existing RNA, but the enzyme will also break down any newly synthesized RNA. Thus, you could not say whether the formation of the cap was due to the expression (translation) of existing RNA or the synthesis (transcription) of new RNA. You should also remember that experiments like this have a basic flaw. By breaking down RNA, you are likely to affect many other aspects of cellular biochemistry. Protein synthesis requires rRNA and tRNA as well as mRNA, so lack of cap formation could be a general effect rather than a specific one resulting from the breakdown of mRNA.

(e) True. The example in the text was the sequence of production of phosphatase enzymes before cap formation. As this sequence still occurs in cells without nuclei, the control over the production of the enzymes must be due to the expression of different mRNA molecules at different times rather than to control over the transcription of the DNA. Thus there must be control over the *translation* of each specific mRNA in the cytoplasm.

SAQ 4 (a) False. As the synthesis of X-ase is occurring anyway, some radioactively labelled protein will be formed irrespective of any increase in the rate of protein synthesis.

(b) False, for reasons similar to those given for (a). Inhibition of protein synthesis would still affect the normal rate of synthesis of X-ase.

(c) True. An increased rate of translation of mRNA for X-ase would give an increased level of X-ase in the cell, provided there was no change in the rate of degradation of this protein.

(d) True—assuming, of course, that the extra mRNA for X-ase was all translated.

SAQ 5 (a) False. The repressor molecule binds to an operator region on the DNA.

(b) False. Lactose induces the synthesis of β-galactosidase by binding to a repressor molecule. The fact that it provides glucose is irrelevant.

(c) True.

(d) False. For example, enzymes involved in metabolic pathways, such as glycolysis, are produced continually. These constitutive enzymes do not require such a control.

(e) False. A promoter is a region on the DNA where RNA polymerase binds.

(f) False. It controls the synthesis of several enzymes and not their structure.

SAQ 6 (a) True.

(b) True (see Figure 25 if you do not understand this).

(c) False. The Britten–Davidson model does provide a way for 'switching on' many operons but so does the Jacob–Monod model (see Figure 23). The crucial difference is that the Britten–Davidson model relies on the activation of specific genes rather than the *repression* of the DNA to allow the production of mRNA.

(d) False. In all organisms, the synthesis of proteins requires ribosomal and transfer RNA as well as mRNA.

SAQ 7 (a) mRNA molecules from different tissues can be distinguished by radioactive labelling. Thus, this is not a good reason to be sceptical of this technique.

(b) mRNA can be separated from tRNA and rRNA, so this is not a reason for doubts.

(c) This is the important factor. Repeated sequences of DNA will affect the rate of DNA–mRNA hybridization. This effect *may* not be crucial, but it does mean that using the data from such studies to support the idea of differential gene transcription may be unwise.

(d) The fact that some tissues have similar mRNA molecules shows only that in *those* tissues the genes have been expressed in the same way rather than differently. If you found that the mRNA molecules were identical in *all* tissues then this would lead you to doubt the idea of differential gene expression. However, if you found a difference in the mRNA molecules between even two tissues this would indicate that genes were being expressed in different ways. The statement, by itself, should not lead you to doubt the evidence provided by competitive DNA–mRNA hybridization.

SAQ 8 (a) Stages II and III, because these curves are very similar.

(b) This must be Stage I mRNA, because this curve shows a rapid hybridization and hence a great number of complementary sequences in the mRNA and mDNA.

(c) You may well have picked stage I, and from what you have learned so far this would be correct. However, because we assume the genetic material, and hence the DNA, is constant in *all* cells you could have taken DNA from *any* stage and still have achieved the same results.

SAQ 9 You should not have found this too taxing as the first part of the Section summary really covers this point! You should have realized, though, that it is only the non-histone protein in the chromatin that seems to regulate gene transcription. This is seen clearly in the experiments on the production of ovalbumin mRNA in the chick oviduct. The role of the histone protein is uncertain—see Section 9.

SAQ 10 (a) Figure 40a shows that there is no hybridization of gastrula mRNA with the DNA. However, from Figure 40b you can see that there is a high level and rapid rate of hybridization. In (b) the DNA–gastrula mRNA sequences must be complementary because there is a high level of hybridization. Thus, graph (b) must have been derived from the hybridization of mRNA and gastrula mDNA. So graph (a) must be based on observations with null mDNA.

(b) Because there is some hybridization with gastrula mRNA in Figure 40b, there must be some tube foot mRNA coded for within the gastrula mDNA. In Figure 40a there is very little reaction with null mDNA. Thus, there can be very little tube foot mRNA coded for in the null DNA fraction.

(c) You can deduce from the graphs of hybridization of tube foot mRNA that most of the tube foot mRNA is already present *at the gastrula stage*. This indicates that there must be control over the *translation* of mRNA because this mRNA, although transcribed in the gastrula stage, is not translated until the sea-urchin is older and develops tube feet; in other words, it is stored.

SAQ 11 (i) The oocyte contains a large amount of 'stored' mRNA.

(ii) At the blastula stage there is a large amount of mRNA present that is not found later in the gastrula. This suggests that proteins may be being synthesized (i.e. this mRNA is being translated) before they are needed.

(iii) More complex stages such as the pluteus have *less* mRNA than the gastrula. Differential gene transcription would suggest that the more complex the organism or tissue, the greater the range of protein needed and hence the *greater* the diversity of mRNA that would be found.

SAQ 12 Yes, if mRNA taken direct from the nucleus is unable to code for ovalbumin in an *in vitro* experiment but mRNA in the cytoplasm can code for ovalbumin, there must be modification of the mRNA between the time it is transcribed and the time when it can be translated. The DNA coding for ovalbumin mRNA has intervening sequences which are transcribed. Thus these must be removed and the remaining mRNA spliced together before it can function.

SAQ 13 (a) Chromatin proteins may affect the way repressor molecules bind to the DNA. However, as DNA is not masked by the chromatin proteins this may not actually affect the action of repressor molecules.

(b) This is not a problem—a repressor–inducer system could work just on those sequences of DNA that do code for mRNA.

(c) This is difficult to explain on the basis of the Jacob–Monod model. The hypothesis would have to be adapted to explain the precise 'tailoring' of mRNA after transcription.

References and further reading

References to the Science Foundation Course

	S101	S100
1	Unit 22, *Physiological Regulation*, Section 3.	Unit 18, *Cells and Organisms*, Section 18.2.
2	Unit 18, *Natural Selection*, Section 1.	Unit 17, *The Genetic Code: Growth and Replication*, Section 17.1.3.
3	Unit 19, *Genetics and Variation*, Section 2.3.	Unit 17, Section 17.12.
4	Unit 25, *DNA, Chromosomes and Growth: Molecular Aspects of Genetics*, Section 7.1.	Unit 17, Section 17.11.
5	Unit 25, Section 8.	Unit 17, Section 17.13.
6	Unit 22, Section 6.2.2.	(no equivalent)
7	Unit 25, Section 4.2.	Unit 17, Section 17.6.
8	Unit 25, Section 8.	Unit 17, Section 17.13.
9	Unit 25, Section 4.	Unit 17, Section 17.9.
10	Unit 18, Section 1.	Unit 17, Section 17.1.
11	Unit 25, Section 4.	Unit 17, Sections 17.8.1 and 17.8.2.
12	Unit 24, *Chemical Reactions in the Cell*, Section 4.	Units 15 and 16, *Cell Dynamics and the Control of Cellular Activity*, Section 15.5.
13	Unit 25, Section 4.	Unit 17, Section 17.9.
14	Unit 25, Section 6.1.	Unit 19, *Evolution by Natural Selection*, Section 19.2.1.

Further reading

The following book provides a good introduction to, and expansion of, the main ideas in the Unit. It could be read in parallel with this text.

Ashworth, J. M. (1973) *Cell Differentiation*, Outline Studies in Biology, Chapman & Hall.

The next recommended text could accompany all the Units on development (11–15); it covers far more material than this Unit. Particularly of interest are chapters on nucleus and cytoplasm (Chapter 1.2), totipotency (Chapter 2.2) and the control of gene expression (Chapter 5.3).

Graham, C. F. and Wareing, P. F. (1976) *The Developmental Biology of Plants and Animals*, Blackwell.

Finally, a book intended for advanced undergraduates which does however put this Unit in perspective and is not too difficult to follow if you have read the Unit first.

Maclean, N. (1977) *The Differentiation of Cells*, Arnold.

Acknowledgement

Figure 10 is reproduced by permission of Oxford University Press from H. Harris (1970) *Cell Fusion*.

List of Units

The Diversity of Organisms

Unit 1	Marine Organisms
Unit 2	From Sea to Land: Plants and Arthropods
Unit 3	From Sea to Land: Vertebrates

Cell Biology

Unit 4	Cell Structure
Unit 5	Macromolecules and Membranes
Unit 6	Enzymes: Specificity and Catalytic Power
Unit 7	Enzymes: Regulation and Control
Unit 8	Metabolism and its Control
Units 9 and 10	Membrane Transport and Energy Exchange

Development

Unit 11	Development: The Component Processes
Units 12 and 13	Cellular Differentiation
Unit 14	Pattern Specification and Morphogenesis
Unit 15	Chicken or Egg?

Animal Physiology

Unit 16	Communication: Nerves and Hormones
Unit 17	Blood Sugar Regulation
Unit 18	Control Mechanisms in Reproduction
Unit 19	Circulatory Systems
Unit 20	Respiratory Mechanisms
Unit 21	Respiratory Gases and Their Transport
Units 22 and 23	Osmoregulation and Excretion
Units 24 and 25	Nutrition, Feeding Mechanisms and Digestion

Plant Physiology

Unit 26	Plant Water Relations
Unit 27	Ion Movement and Phloem Transport
Unit 28	Plants and Energy
Unit 29	Plant Cells: Growth and Differentiation
Units 30 and 31	Morphogenesis in Flowering Plants